Managing Geeks

A Journey of Leading by Doing

By
Andy Leonard

©**Andy Leonard**
All Rights Reserved 2014

ISBN: 1502947609

COVER ART BY TODD RADNEY – YESTERDAYS PHOTOGRAPHY

Table of Contents

About the Author — 9
About this Book — 9
Introduction — 10
  Chapter 1. Human Doings — 16
Management Fads — 16
What Do You Do? — 16
Who Are You? — 17
It's Business — 17
Do What You Are — 18
  Chapter 2. The Integrity Challenge — 19
  Chapter 3. Good Will: Negative and Positive — 20
What Do You Stand For—Or Stand Against? — 20
Stand Against What? — 20
Unfairness and Bad Will (Negative Good Will) — 21
The True Cost of Bad Will — 23
How to Fix Bad Will — 24
Good Will, Finally — 24
  Chapter 4. Visions, Quests, and Missions — 26
The Vision, After the People — 26
The Quest — 27
The Mission — 27
The Right People — 28
  Chapter 5. Right, Wrong, and Style — 30
Style — 30
Experience — 30
Really, It's No Imposition — 31
Disagreement Is Good — 32
  Chapter 6. Follow Me — 34
Go Do That — 34
Follow Me — 35
  Chapter 7. Balance — 37
Balance — 38

Balance and Transparency	39
Balanced Transparency and Leadership	39

Chapter 8. Workplace Hierarchy — 42

Compensation	43
Job Security	44
Esprit de Corps	44
Organizational Respect	45
A Kick-Butt Team	46

Chapter 9. The 15-Minute Meeting — 47

Multitasking	47
Time Slicing	48
The Beauty of the 15-Minute Meeting	48

Chapter 10. Problems — 50

The Effects of Problems	50
Drama Is More Work	51
Some Irony	51
Overcoming Drama	51
Terror	52
A Scenario	52
False Assumptions	52
The Unexpected Isn't Supposed to Happen	53
Every Primary Resource Has a Backup Resource	54
When Terror Sets In	55
Terror Makes More Work	56
Some Irony	56
Overcoming Terror	56
The Right Question	57
Process Improvement	58

Chapter 11. Software Is Organic — 60

Listen to the Developers	62
Battle of the Three Letter Acronyms (TLAs): WBS vs. EBM	63
Organic Software Management	64
Conducive Conditions	65
Right Stuff, Right Time	66
Taking Care of the Team	67

Chapter 12. I Don't Work on My Car But I Still Work on My Truck ... 68
Things Change ... 68
Yes and No ... 68
"I Used To Be a Developer" ... 69
Odder Still ... 69

Chapter 13. A Turning Point ... 71
Secrets to Happiness ... 71
A Large Shovel ... 72
Does This Work? ... 73
All the Time? ... 73

Chapter 14. Now if You'll Excuse me, I'm Going to Go Screw Something Up ... 74
Crap Happens All the Time ... 74
This Is About the Future ... 76
Enter Experience ... 77

Chapter 15. Getting It Right the First Time ... 78
Process ... 79

Chapter 16. One-Time Productivity Boosts ... 81
The Cause of the Effect ... 81
Turn It Up ... 82
The Value of Change ... 82
The Central Rule for Projects™ ... 83

Chapter 17. Institutionalized: A Geek Tragedy ... 84
Reacting ... 84
Institution ... 85
What Should Have Happened ... 85
Does This Still Happen? ... 86
What's the Big Deal? ... 86

Chapter 18. Perfection versus Precision ... 88
Semantics ... 89

Chapter 19. Diversity ... 90

Chapter 20. Ringing ... 92
Code Churn ... 92

Bug Rate and Severity ... 93
Developer Enthusiasm ... 93
Chapter 21. Passion ... 95
In Community ... 95
Salary? ... 97
Chapter 22. Business Losses and the Power of "I Don't Know" ... 98
Chapter 23. Personality Clashes, Style Collisions, and Differences of Opinion ... 101
One Solution ... 102
Chapter 24. Human Resources ... 103
Human Resources Is Dehumanizing ... 103
HR Flattens ... 104
Human Resources Is a Motivation Sink ... 105
Not Just Money ... 105
Reward Doesn't Happen in a Vacuum ... 108
Chapter 25. Sounds Good ... 109
Unintended Management Consequences ... 110
Patterns vs. Exceptions ... 110
Serving ... 111
Chapter 26. Coopetition ... 113
How You Play the Game ... 113
Short Term ... 113
Long Term ... 114
Sustainable ... 114
There's Enough Work to Go Around ... 115
Chapter 27. Disruption ... 116
Change Hurts ... 116
Chapter 28. Recognition versus Satisfaction ... 119
Get. Started. Now. ... 120
Chapter 29. Do You Have a Job, or Does Your Job Have You? ... 121
Why These Questions Matter to You ... 122
Chapter 30. Conviction ... 124

Firmly Held Belief or Opinion	124
Firmness of Belief	125
Balance	125
Courage	126
Chapter 31. I Type in Real Time During Demos	**127**
The Positives	127
Chapter 32. Techganic	**129**
Trampled	130
Fast Growing and Withering	131
Choked	131
Successful	132
Chapter 33. The Playing Field Is on a Hillside	**133**
Chapter 34. How Is Your Serve?	**135**
Chapter 35. Volatility	**136**
Strategy	136
Responding Strategically	137
Volatility	137
The Cure	138
Chapter 36. Engines of Loss and Gain	**139**
Solution (and Anti-Solution)	140
Increase the Odds	141
Chapter 37. Associate of Applied Science	**143**
Chapter 38. Freedom to Innovate	**145**
Blocking Requests	146
An Example	147
One Solution	148
We Can All Benefit	149
Chapter 39. Love Your Opponents	**150**
Why You Should Love Competitors	150
Chapter 40. Outlasting Outrageous Opposition	**152**
Chapter 41. Ensuring that a Metric is Effective	**154**
Change the Metric	154
Eliminate the Metric	155
Fire Those Responsible for Gaming the System	156

A Better Solution	156
Chapter 42. Evil Is Easy. Creating Is Hard	**158**
Chapter 43. Performance-based Management Doesn't Work	**160**
My Question	162
Chapter 44. Data Visualization and Dashboards	**163**
Simple Is Good	163
Data Visualization Can Be Dangerous	165
Elegant! Equals Pretty	167
Communicate	168
Chapter 45. Credibility	**169**
Credibility and Values	170
Interpretation	170
What I Say Is Important	171
What I Do Is Important	173
Chapter 46. To Snark or not to Snark	**175**
The Solution	176
Chapter 47. Question	**177**
Chapter 48. Managing Confidence	**178**
Managing Confidence	179
Chapter 49. Push the Pebble	**181**
Dams and Avalanches	181
One Pebble	181
Kick the Pebble	182
Chapter 50. Less-Useful Soft Skills	**183**
Free Advice	183
Retentiveness	183
Destructive Competition	184
Chapter 51. Can You See Me think?	**185**
Measuring Manufacturing Work	185
Measuring Intellectual Work	186
Divide and Conquer?	186
The Point?	188

www.linchpinpress.com

About the Author

Andy Leonard is CSO of Linchpin People, an SSIS Trainer and Consultant, SQL Server database and Integration Services developer, SQL Server data warehouse developer, community mentor, blogger, and engineer. He is a co-author of SQL Server MVP Deep Dives, Volume 2 and SSIS Design Patterns. His background includes Visual Basic and web application architecture and development and SQL Server 2000-2014. Andy is one of the founders of Linchpin People®, and loves SQL server work. You will often hear Andy talking about what a blessing it is to be able to help his customers solve their database problems.

Introduction

My independent consulting career ended in mid-2008. At least I thought it had ended. The US economy was suffering, and the demand for expensive consultants had dropped. I sought full-time employment and began by making a list of former clients I found most intriguing. At the top of that list was Unisys.

That's not accurate. At the top of that list was a person, Bennett McEwan, not a company.

Ben worked for Unisys. He was a friend of Brian Moran, CEO of the consulting company I was contracting through. A year earlier, Brian had sent me to Unisys to teach Ben's team SQL Server Integration Services (SSIS). Ben and I struck up an immediate and lasting friendship.

When I met Ben I did not realize I was speaking to the person who had taught me Transact-SQL (T-SQL, the declarative programming language used by Microsoft SQL Server). Ben wrote *Teach Yourself Transact-SQL in 21 Days*, a Sams publication. Ben wrote more books, and I had written a few books when I met Ben. He had made the transition from independent consultant and author to the corporate world.

When I accepted the offer to work for Ben, he said, "I will coach you on management. I think you will enjoy it and be good at it."

I responded, "I think you're wrong on both counts."

Despite my objections, Ben assigned a small team of four database developers to me. After a couple months he asked me, "So how do you like management?" I told him I didn't like it at all. I didn't feel competent and I still didn't want the job. I concluded with, "I didn't come ask you, 'Hey Ben, can I be a manager?'"

I liked being a lone gun. I felt competent serving as a data integration architect for the Extract, Transform, Load (ETL) team, building solutions using SSIS. It's a good feeling, the feeling that you know what you're doing. I didn't feel that way about management at all.

A few months passed, and the ETL team manager position became open. Ben called and asked if I would manage this much larger team. I thought about it for less than a minute and accepted.

Did I feel better about my management skills? Not at all. I had been serving the ETL team for months as architect and knew many of them well. I knew some from before when Brian Moran had originally sent me to train Ben's ETL team. I liked them. I reasoned I could manage them because of that.

I was right.

Over the eighteen months I served as ETL team manager, the ETL team varied in size from about 28 to 40 individuals. Some were full-time, others were contractors, and some of the contractors were offshore. Ben continued to invest time and wisdom in me. He also provided much-needed air cover when I made mistakes.

I made lots of mistakes. Learning from one's mistakes is at once the best and most difficult way to learn. Ben was patient and an excellent coach. I learned about managing technical teams from him. I received awesome—though sometimes painful—advice and feedback from him.

I look back on my two years working for Ben McEwan as the best two years I spent working for anyone anywhere. That is why I dedicate this book to *Ben, my mentor, coach, and friend.*

As mentioned previously, Brian Moran was CEO of a consulting company I was affiliated with from 2006-2008. Along with almost everyone else interested in learning SQL Server, I had been reading Brian's articles and editorials in *SQL Server Pro* magazine (http://sqlmag.com) for years.

I first met Brian in person when I delivered a presentation about SSIS to the Northern Virginia SQL Server User Group (http://www.novasql.com). This was my second SSIS presentation delivered after my first book (*Professional SSIS 2005*, Wrox, 2006, http://www.amazon.com/Professional-Server-2005-Integration-Services-ebook/dp/B000VIBWMG) was published.

I must digress a little and discuss my *first* SSIS presentation after the book was released. The important part: Three people were in attendance, counting me.

Yep, such is the life of a big shot rock star author. We actually had a great chat about the capabilities of SSIS, and then I drove the three and a half hours back home, got to sleep around 1:00 AM, and got up four hours later to drive to work. I was not bitter, though. I promise. But I had doubts about the popularity of me and the popularity of SSIS. At that point, I didn't know who or what was to blame for not drawing a larger audience, but I strongly suspected it was me.

The second presentation was scheduled for a month later, and I didn't really think about it. I also didn't really prepare for the presentation. After all, I expected a similar turnout.

I was wrong about the turnout. Way wrong. There were 125 people there, including Erik Veerman, one of my data integration heroes and co-author of several books on SSIS (http://www.amazon.com/Erik-Veerman/e/B001IGNQKG).

The group leader was… animated. He gave an introduction that made me sound smarter than I was (or am). I think my first impression of the group leader was, "He sure has a lot of pep and energy." Compared to me, a lot of people have more pep and energy.

I started the presentation, said my piece in 20 to 25 minutes, and asked if there were any questions. There were a *lot* of questions. I did not know the answer to most of them. Partially because I was woefully unprepared to present and partially because I, too, was relatively new to SSIS. I understood the parts of SSIS I had written about, but I'd only written two chapters in the book!

I was in way over my head.

Thinking on my feet, I began asking members of the audience if they knew the answer to those questions. Erik (God bless him!) was able to answer many and (God bless him, again!) didn't call me out for delivering a presentation on a topic without knowing more first.

As the meeting wrapped up and I was feeling more stupid than ever, the group leader handed me his business card. "Brian Moran," it read, "CEO" of a company I would really have liked to be affiliated with if only I hadn't blown this presentation right in front of him.

Fast-forward six months: Brian called me and said he wanted us to work together. I was floored and excited—all giggly inside actually—and one of my questions was, "Why me?"

Brian told me it was because of that presentation six months earlier.

I said, "But I blew that presentation."

"Yes. Yes, you did," Brian lovingly responded, "But you didn't run out of the room. Instead you hung in there, switched gears, and everyone learned stuff about SSIS they didn't know when they walked in."

I think I responded intelligently along the lines of, "Huh."

Brian continued, "We can teach you technical stuff you don't know. We can put you on gigs with others who will coach you. But we can't teach you to think on your feet like that." And so I was in.

I didn't get to work with Brian much or often. And, truth be told, I was more of a practicing redneck when we first collaborated—which is to say I was hard-headed and trifling. I did not know soft consulting skills existed, and it showed. I am an engineer by training and, really, by birth. I didn't learn until later in life that I suffer from Attention Deficit Hyperactivity Disorder (ADHD), but it was pretty obvious to everyone except me.

Brian tried and had some success at gentling my soul. I learned a lot from him each time we interacted. I admired him greatly, which helped me learn from him. I still struggle with learning from people I do not admire. And, I still admire Brian. Actually, I admire him even more today than I did back then.

Part of the reason I admire Brian is I have been his business partner for the past three years. I have had the privilege of getting to know him much better. And I like him anyway! Kidding aside, I have come to respect Brian's gifts and his heart that is so willing to share his gifts.

What are Brian's gifts? He is a visionary. He has a mind capable of multi-dimensional thought. He loves to help people. He enjoys solving problems. Combined, these gifts compel Brian to see multiple ways to solve problems for individuals and businesses. I love just being around him, just talking with him. To me he is fascinating and I just like watching him go!

He is my brother in the Christian faith. Brian has continued to coach me where Ben left off. I am blessed to be Brian's business partner and honored to be counted among his friends.

That is why I dedicate this book to *Brian, my brother, business partner, coach, and friend.*

What follows began as a series of blog posts. I began sharing my journey into management at SQLBlog.com in December 2009, roughly one year after starting to manage the smaller team of database developers at Unisys. I have (mostly) maintained a habit of not blogging about occurrences until at least one year has passed. Why? Perspective. Time and experience have a way of shaping the past with context, while at the same time defusing any present tension.

To make sense of this collection of islands of my blathering, Linchpin Press engaged the services of a most gifted editor: Karen Forster. Trust me, Karen did awesome work. If things flow and make sense to you, dear reader, you have Karen to thank. If things seem disjointed still, you can blame me and rest assured Karen did everything in her power to bring harmony.

Andy Leonard
Farmville, Virginia, US
July, 2014

Chapter 1. Human Doings

My Granny used to say "Pick a way to be and be that way." I've interpreted and applied that, and I say:
> *Be who you are.*
> *Do what you are.*

This chapter and this theme of this book are about being as opposed to doing. Experience has taught me that to be a good manager, I need to know who I am and do what I am. This book reflects the journey that produced that experience.

Management Fads

Many of us have lived through the management fads of recent years. They range from dressing for success, to seven habits, to elevator pitches. I'm not knocking these fads (well, not much anyway). There's good stuff here. But what strikes me about all these fads is that they tell you what you should *do*, not how you should *be* to become successful. I have learned that success as an employee, and even more as a manager, is not about what you do. It's about who you are. That's the foundation of my career, and the focus of this book.

Anyone who's seen me knows I could stand to dress better. I have more than seven bad habits. I chew up a good 10 seconds saying the word "uhhh" during any attempt at a two-minute elevator pitch. But will doing things to change these attributes make me successful? They might help, but I maintain that I'll still be what I am under the new clothes, good habits, and succinct communication. What I can accomplish as a manager depends on the kind of person I am and what values I hold. Success is about being, not about doing.

What Do You Do?

I sometimes struggle with answering this question. When my neighbors ask me what I do, I tell them I work with computers. My neighbors are really good people, but they're not in the technology industry. Launching into a well-rehearsed and concise lecture about technology really isn't going to impress them. Well, I take that back: It likely will impress them, but not in a positive way.

I don't have an elevator pitch for what I do. And I don't want one. The problem is the question. What I do is not important. What I am is what matters.

I am a human being. I am not a resource. I don't have an elevator pitch because I don't want to sell you anything—especially not myself.

If you want to rent some of the time I've been given (my definition of "work"), we can make that happen. If you want to discuss something, I'm in. If you share a passion for technology, good engineering, and great teams, odds are we'll learn stuff from each other.

And that's the heart of it for me: learning. Not selling. There's a difference. One key is to ask the right question, which I believe is, Who are you?

Who Are You?

I could show you my business card. A lot of thought went into this card. My card does not communicate what I do. It communicates who I am: a human being, not a human doing. I'm a package deal. If you need a person with SSIS experience, you also get a Southern Boy at no extra charge.

It's Business

Here's the distinction: I am Andy; I do work. I don't mix business life with personal life. I do business personally. That's what I want my business card to convey. I believe it does. And I believe that finding out what that means—in terms of my values, my ethics, my relationships, my personality—is what makes me a good manager. What I am is what I have to offer the world, and it's the core of what I have to offer others who want to be good managers.

Do What You Are

As Buck Woody reminded me in his blog post, *No Certainties, Including That One*: I am not relating "some newfound wisdom as if I'm the first one to discover it." At least I hope I'm not sounding like I think this is all new. It isn't. The stuff in this book is stuff I've learned from others.

That's why this book is a recommendation that you do what you are. All of us are built for something. Figure out what that something is, and do that. It may be in your chosen field, it may not.

I am not saying, "Do what you can," because you can do lots of things, and perhaps do them very well. The question is, do those things fulfill you? If you had a day with absolutely nothing to do, would you go do those things?

Passion is a clue to what you should do. If you're not doing what you are, stop doing it. Start being.

I'm not advocating you walk out of the front door and quit your job this instant. I am seriously advocating you begin searching for a position (or start creating one) that matches who you are.

If you're a human doing, stop. Become a human being!

Chapter 2. The Integrity Challenge

This chapter is a challenge to all who read it. For as long as it takes you to read this book and for the next month, I challenge you to be utterly honest with yourself and with those you interact with. This includes those who make requests of you, those of whom you make requests, friends who IM, folks who call—everyone.

I'm not challenging you to be transparent. You do not have to tell them everything, but, if you accept this challenge, you do have to tell them enough so that the truth is communicated.

If you decide to ditch an appointment because you learn of some more-fun-thing to do, you need to communicate that you'd rather do the more-fun-thing to the person expecting you to attend the less-fun-thing. If you accepted a side gig and a better-paying side gig appears, you need to communicate the truth to the person you made the original agreement with.

For one month, put the politics aside. You don't have to be mean to be honest, so don't be mean. Just be honest.

If you accept this challenge, you will learn things about yourself, your friends, and those with whom you interact. It will also help you understand the point of this book: To be a good manager, you have to know yourself and your values.

Chapter 3. Good Will: Negative and Positive

In business and in managing teams, every action has consequences. Generally, good actions have good consequences, and bad actions have bad consequences. That may sound like a platitude, but success is built on good will. It's important to keep that thought at the forefront of your approach to life and to business. Good will is good business.

What Do You Stand For—Or Stand Against?

A friend and I shared a recent IM conversation, and she told me one of the two netbooks she had ordered had been stolen in transit. They would have been gifts for her children, and one gift was gone—within two weeks before Christmas.

Tracking numbers showed the netbook had been stolen while in possession of the shipping company. Most likely, an inside job.

This story got me thinking about the mindset of an individual who would risk job, reputation, career, and potentially freedom for the promise of some extra cash. What does such a person stand for? What boundaries does such a person have? Are there limits to what behavior and actions that person would find acceptable?

Those questions made me consider what I stand for. I realized what I stand for is largely defined by what I stand against, both as a person and as a manager.

Stand Against What?

A key issue is that I need to be clear about what I stand against so that people don't misunderstand my position. Let me give an example: Because of some comments I've made about project managers and MBAs, some folks have formed the impression that I'm anti-PM or anti-MBA. Nothing could be further from the truth.

I'm anti-stupidity. This does not imply that I'm calling PMs and MBAs stupid. Nor does it imply any bad will towards anyone. Standing against stupidity first means defining it. Understanding stupidity lets me find ways to generate good will.

Actions I would initially define as stupid may not be so. Stupidity, then, bears testing to see if it's really stupidity or something else. If it is truly stupidity, I need to find a way to stop the stupidity and solve the problem. If it isn't really stupidity, I can get to the root of what's making something look stupid and address the cause before bad will results.

As a manager, stupidity can bite in a several ways:

1. If I condone stupidity, the work and team morale will suffer.
2. If I mislabel something as stupid, an individual will likely suffer, which can lead to the work and team morale suffering.
3. If I am stupid, I and others will suffer for it.

That's why I need to examine all aspects of the problem, including the people involved, and make sure I understand what's going on so that I can deal with it in a way that creates good will and removes any bad will that has arisen.

Unfairness and Bad Will (Negative Good Will)

I also stand against unfairness. A lot of unfairness falls into the stupidity category and is captured by my filter that checks for stupidity and deals with it.

Prominent, respectable people and organizations do dumb things sometimes. I doubt seriously it's intentional, and it doesn't start out as all-grown-up-evil-unfairness-and-stupidity. It grows into that from a tiny seed of unfairness, greed, or innocent competition, or decision fatigue, and that can blow up into negative good will, or "bad will," the evil twin of good will.

As a manager, you can find endless instances of unfairness. Dealing with unfairness means finding a way to do what's right for everyone. If you deal directly with unfairness, you can avoid bad will.

One way to derail unfairness is give and take. I've said it a hundred times during negotiations: If you're going to bring a stick, have the decency to show up with a carrot. Translation: If you're going to ask for something from someone, offer something else in return. Why? Because although you can stir up a swarm with any old pile of crap, you still catch more flies with honey.

Making sure everyone gets something leads to trust and respect, which go a long way in business. You have to earn respect, and that respect is directly proportionate to how well you support others and how well you get the job done. Stupidity or unfairness will not be tolerated, so as a manager, you need to make sure that you are not the source of stupidity or unfairness and that you do not let stupidity or unfairness go unchecked. You need to make sure that everyone is doing good work and not contributing to stupidity and unfairness.

This point is well made in the context of IT by J. Ello's excellent opinion piece: "The unspoken truth about managing geeks":

> *"Few people notice this, but for IT groups respect is the currency of the realm. IT pros do not squander this currency. Those whom they do not believe are worthy of their respect might instead be treated to professional courtesy, a friendly demeanor or the acceptance of authority. Gaining respect is not a matter of being the boss and has nothing to do with being likeable or sociable; whether you talk, eat or smell right; or any measure that isn't directly related to the work. The amount of respect an IT pro pays someone is a measure of how tolerable that person is when it comes to getting things done, including the elegance and practicality of his solutions and suggestions. IT pros always and without fail, quietly self-organize around those who make the work*

> *easier, while shunning those who make the work harder, independent of the organizational chart."*

Oh, and don't think that unfairness happens just between employees and employers. It also happens to independent contractors and larger corporations. The last time I checked, a contract was an agreement between two (count 'em, two) parties. Just because you're a big company, you do not have the right to push around the little guy. In fact, that's a recipe for bad will, and bad will has a cost.

The True Cost of Bad Will

The true cost of bad will (which is really just negative good will): Your team will have no trust, and respect will be conditional. As the quote above from J. Ello's article makes clear: Trust and respect, based on everyone doing good work, are at the core of the business of managing people in technology. And I think that applies to any industry.

How long does bad will live? It outlives and outlasts all the benefits previously gained by the person or company before someone behaved poorly. Even if it's an isolated act (and such things usually are not isolated acts, but indicators of a pattern of behavior), that one act remains in the memory of all who witness it or learn of it. Well beyond the scope, benefits, and immediate consequences, bad will becomes the reputational equivalent of a criminal record.

Thank about it. Why does the justice system keep criminal records? Because everyone has an internal list of acceptable actions. Criminals are people who have demonstrated they're capable of breaking the law to obtain what they want.

The law isn't a barrier to their behavior; it doesn't play into the equation of their acceptable actions. They've done it once, and they're capable of doing it again. So information about their modus operandi and personal information about the individual (e.g., fingerprints, DNA) is kept on file by the authorities. If a similar crime is committed, the authorities start with people they know have committed this act in the past.

The idea of the repeat offender also applies to personal and corporate reputations. If individuals or companies treat another individual or company unfairly, they are demonstrating that, in their minds, treating people unfairly is on their internal list of acceptable actions. This leads others to expect bad behavior.

When it comes to dealing with folks who've mistreated others, this thought is foremost in everyone's mind: "If those people or companies did something bad to someone, they could do it to me."

How to Fix Bad Will

You want to know how to avoid bad will? Simple. Treat others as you would have them treat you.

"Great. But what if I've already done the deed?"

If you've abused a relationship, there's a model for that (taken from the same book as the last sentence in the first paragraph in this section), as well. If someone compels you to give them your coat, give them your cloak also. If someone asks you to walk a mile with them, walk two. In other words, you make it right by more-than making up for the original mistreatment.

"That sounds expensive."

It is. It costs more than simply treating others as you'd like to be treated in the first place, but it costs less than years or decades of bad will.

Good Will, Finally

Everyone makes mistakes. People err. Contractual parties disagree. It happens. And it will continue to happen.

But a funny thing happens when you do business as I've described above: Trust grows, respect increases. Before you know it, you suddenly have more business than you imagined. People and companies come to you, and they come to you first.

I know some companies and individuals who conduct business in this manner. As mentioned previously, they are not flawless nor faultless. They do, however, enjoy the most respect and highest credibility in their industries.

This is good will in action. And in the bank account.

Chapter 4.   Visions, Quests, and Missions

People come first. In his excellent book, <u>Good to Great</u>, Jim Collins makes the point that you start with the right people, and then go. That seems backwards to some. Shouldn't the vision come first, and then you go find the right people? No. Not according to Collins, and not according to my experience.

"How does it work if you don't have a vision to guide the team?"

Glad you asked. A lot of evidence says it really comes down to the team, and the vision grows from the people. The premise is: Get the right people together and a good, well-executed idea will fall out of the mix.

What about a business plan? What about all the other stuff you're supposed to do before building a team? You need the right people with different types of skills that can take the team from a vision to a quest that accomplishes your mission.

The Vision, After the People

As you're putting together the right people, be aware that one of them (maybe you?) needs to be a visionary. Visions require a visionary. Visionaries are important to a business or project because they have a big picture in mind. They are usually good communicators, and they articulate the vision in a way that inspires the team.

When you're looking for the right person for this role, remember that some visionaries have few or no other skills. This doesn't decrease their value as the keeper of the vision; it just means you keep them away from the other work!

Not all visionaries are hands-off. In the technology world, I've worked with some stellar architects who are visionaries and great designers. Those folks are rare, but they exist.

The Quest

There are lots of comparisons of project teams to people on a quest. Why? You're rarely asked to do the same thing twice in the software business. You always have some challenge, some obstacle to overcome, something new to aspire to. The comparison to a quest can help you determine which people are best suited to form the team responsible for going to whatever ends it takes to accept the challenges, overcome the obstacles, and chase the next new quest.

Quests share a couple characteristics: They are difficult and are undertaken by heroes. The ideal you're looking for is a team of individuals who are not afraid to be heroes, who are guided by a vision of the end product, and who work on a project with enthusiasm and a sense of purpose. People on a quest are often consumed by the need to reach the goal. You want the sense of dedication you see in a knight on a quest.

The Mission

A lot of businesses develop a mission statement as a first step in establishing goals or correcting their course. A mission statement isn't a bad thing; it just doesn't belong at the top of the to-do list.

What belongs at the top of the top of the to-do list?

People. See above.

Simon Sinek, author of Start With Why (Penguin Group, 2009), observes, "Most companies' vision and mission statements are useless."

I think what Sinek is getting at is that something *beyond* making money (especially making money for stockholders) must drive the visionary and the team. That "something beyond" is the mission itself—*not* the mission *statement*. The mission is what the quest is seeking to attain. The mission focuses the quest so that the team is all moving towards accomplishing that mission.

A mission is different from a mission statement. Mission statements are expressed and executed over time by groups. A mission is focused on a singular outcome. Mission statements are undertaken by groups; a missionary conducts a mission. You need people who will take on the mission and be missionaries who go out and line up whatever support they need to ensure that the mission gets done.

The Right People

With the vision, quest, and mission in mind, you need to make sure you have all the right skills and personalities on your team. Get the right people on the team and the wrong people off the team.

That second part may sound harsh, but think about it: You're not really doing anyone a favor by forcing the wrong people to participate on a team where they don't fit.

And please listen: If you don't fit on a team, that is not a mortal sin. It shouldn't be counted against you. You came, you saw, you worked hard, and it just didn't work out. That kind of stuff happens to everyone.

Everyone?

Yep. Everyone.

In my experience, the order of operations for a successful team is:
    1. People
    2. Vision / Quest
    3. Mission

A vision can be extremely well conceived and inspiring. A mission statement can capture where the team or organization is headed and provide direction. But people are the key to bringing it all together and actually accomplishing the vision and the mission. As a leader, it's your responsibility to put your people first and to ensure that the right people are in the right place. Once you have the people, the rest will fall into place.

Chapter 5.   Right, Wrong, and Style

I beat on the right-and-wrong drum a bunch in this book. Clearly, some things conveniently and neatly fall into these categories. And some things do not. A lot of those latter things fall under the heading of style.

Style

Style is about how you tackle something when there's no objective right or wrong way to do it. When right and wrong are not crystal clear, people have different ways of determining what to do.

Everybody has her own style, developed out of her own experience and the environment she find herself in. Experience is always a good teacher, and you can proceed according to what has or hasn't worked in the past.

Another option is for a person in authority to just mandate that a certain way will be followed. Another approach is allowing different people to try different ideas. Disagreement, as long as it's constructive, can enable teams to come together and establish consensus. I think these styles are valid, and I use them.

Experience

I manage people. The people I manage actually manage people. And the people they manage develop SQL Server Extract, Transform, and Load (ETL) processes. Now, I've written ETL processes since before I knew they were called ETL. That doesn't make me an expert, but it does make me experienced. I have my own way of approaching solutions, my own style, based on my experience.

Experience doesn't mean I know everything. You can check with my team. They'll tell you that I don't know everything. In fact, experience has shown me it's best to keep an open mind and learn something new every day.

That's why I regularly tell members of my team: If you see something that looks dumb, email or IM me and say, "Andy, this looks dumb." I will respond in one of two ways:

1. "It's designed that way because..."
2. "You're right, that is dumb."

The result of this dialog will be a better product for our customers.

Experience has taught me to learn from others and to learn from mistakes. As a practical example from the technology world, one reason I love test-driven development (TDD) is because it's built around failing first. I fail all the time. When I found a methodology that incorporated failure into its cycle, I was home. TDD fit my style.

Really, It's No Imposition

Now I could mandate everyone on the team do things that fit my style. But that would be dumb. Why? Because (brace yourself) people are different. Imposing my style on the team would not improve things—it would hamper innovation.

This is why rigid standards usually fail outright or are ignored. So what's the right way to convey best practices?

It depends. The answer is different for different leaders, because there are different styles of leadership, as well. I make heavy use of something I call the Buffet Model. This model isn't named after a rich Nebraskan (his name has two Ts); this is from the salad bar type of buffet.

Basically, I demonstrate the best way I know to solve a problem or address an issue. If that works for you, you are free to use it. If not, you can keep doing it your way so long as it doesn't really mess up the rest of the effort.

At my buffet, I believe if you see me enjoying the stuff on my plate, you will say, "Hey, that looks good. What is it?" You might then decide to opt for steak instead of beef broth.

There are good reasons for sticking with the broth in some instances, so we don't make declarations at Andy's Buffet like, "Broth stinks. Always have the steak." What we do instead is offer two dishes for obtaining beef protein on the buffet, and you can choose which one to consume.

Before this analogy is stretched beyond usefulness, let me stop here. My team consists of software developers, so the "buffet selections" are actually design patterns. What's more, any developer on the team can add to the collection of design patterns at any time.

So, what happens if a developer insists on developing in a style that really messes up the rest of the effort? We talk about it. The goal of these meetings is to learn from each other. Each side presents its arguments, and the goal is consensus. Most of the time, we reach consensus and choose a pattern that works best. Sometimes we decide both patterns have merit, and we agree on when to use each. On rare occasions, this meeting deteriorates into something akin to an intervention. And every once in a blue moon, I have to be the heavy and say, "Please do it this way for now."

Disagreement Is Good

When I was in the Virginia Army National Guard, I learned about overlapping fields of fire. It's part of a foxhole perimeter defense tactic that allows me to take cover behind a berm and shoot at the people running and gunning towards you, while you take cover behind your berm and shoot at the people running and gunning towards me. Keeping you safe keeps me safe. We win together.

What does this look like in the software workplace? I find an apt analogy when I work with people who have different styles. They are going to look at any business problem from their perspective, and it's different than mine. We are going to disagree. They may even say "Andy, this looks dumb."

So long as there is trust and respect, we can cooperate—even if we have different styles or disagree. This is the first step to a better solution. In the end, we win together.

Right and wrong are generally clear cut. Deep down, we all know the difference and recognize which is which. But style is where differences come into play. By recognizing and allowing differences in style, a leader can gain in many ways, including gaining the respect of the team. There's a difference between "wrong" and "different." This lesson is the start to understanding the true power of diversity.

Chapter 6.   Follow Me

> "Wars may be fought by weapons, but they are won by men. It is the spirit of the men who follow and the man who leads that gains victory." - General George S. Patton, Jr.

That quote from General Patton gets to the point I've been making: It's all about the people and the visions, missions, and quest they pursue. The spirit Patton refers to is what I'm getting at when I talk about right and wrong, goodwill and bad will, and style.

A leader's job is to get the team to accomplish its task. Some leaders like to motivate their teams by telling them what to do. Others like to inspire. I've experienced both types of leadership first hand. Because I believe that the people come first, I prefer the type of leadership that inspires.

Go Do That

I used to work for a living. Before joining the software community, I held a bunch of different jobs. I've been a truck driver, peach picker, tobacco puller, soldier (part-time), factory laborer and electrician, electronics technician, electrical engineer, and farm hand—to name a few. Most of those jobs involved work.

Nowadays, I think and type. There's a difference. Work hurt my back and sometimes my arms and legs, too.

A lot of my bosses were go-do-that bosses. They would point to something—a row of tobacco, a stand of peach trees, a pile of manure, the perfect location for a foxhole—and say something like "Leonard! Take care of that."

It was my job to go do the work. It was their job to tell me which work needed to be done. It wasn't very collaborative, this go-do-that style of leadership. But some quests and missions require this style of leadership.

Follow Me

Other bosses jumped into the trenches (or code, or foxholes) with those they led. These were follow-me leaders. They got out front. They physically led. If I worked on a weekend, these bosses would usually show up too—not every weekend and not for the whole day—but they made an appearance and gave up some of their time alongside the team.

I think this style of leadership works better. Why? Nothing builds loyalty like a leader out front. And when the team (or squad, or family) is out of magic—when the deadline looms or more enemy troops than expected show up—you need something beyond talent or training. Loyalty will help.

I find follow-me leadership skips right past motivating people and into inspiring them. As Patton said, the people and their spirit are what makes a winning team. The follow-me leadership style respects the spirit of the people and instills loyalty better than anything I know. It's inspiration instead of motivation, and inspiration will always take you farther.

The Tools of Inspiration

You can never tell where inspiration will come from or what resources will end up inspiring you over and over through the years. The Patton quote that opens this chapter came from what you might think of as an unexpected source.

I still have a tattered copy of *Leadership and Command at Senior Levels*, the one I received as an attendee of the Virginia Army National Guard Non-Commissioned Officer Academy. Most leadership books aren't a matter of life and death, or the future of a nation. This one is about those things.

Each chapter starts with a quote from a military leader. You may be surprised to find concepts that apply to leading a software project or any business project in these pages (or you may not be surprised; I suppose it depends on your background).

That old military book is chock full of leadership goodies that can be applied by all people who find themselves in a leadership role: managers, directors, user group leadership, team leads, and consultants in a project or technical lead role.

What this book has taught me is to look for leadership guidance everywhere. Don't dismiss an unlikely source of inspiration. You never know what you'll find.

Chapter 7.   Balance

Be the best you can be. Sometimes everyone feels that work is getting in the way of life. The hard work you do to make your "real" life possible can leave you feeling guilty and that you're neglecting what's important to you.

But there's another way of looking at the balance between your work and life. Want to know what's important in your life today? Look at your plans for the day. Don't just look at your Outlook calendar. Look at the plans you have for the evening and the morning before work. The things that are important to you are there.

You have to include things that you normally don't think about: You're going to call your spouse or significant other just before you eat lunch to see how their day is going, right? That's important. When you finish working this afternoon / evening / morning, you have a standing date with your kids to play a video game, or help them with their homework, or just hang out with them and watch television, right? That's important. You and your family participate in weekly activities, as well—stuff like soccer practice or karate or dance class (my daughter Emma calls hers "dance cwass") or religious ceremonies or meetings on the weekend, right? That's all important. On your professional side, you support monthly, semi-annual, or annual developer community events or code camps or SQL Saturdays by attending, volunteering, or presenting, right? That's important.

It's easy to feel guilty about working too much. It's really easy in IT. I know lots of geeks who feel guilty about working too much. I don't know anyone in this category who can afford not to work. If you have enough money in the bank to handle or insure every foreseeable future event and you're still working too much, then you should feel guilty. For the rest of us, work is necessary.

Notice I didn't say it's a necessary *evil*. If your work is evil, stop right now and find another job.

If you're working to support yourself and your family, stop feeling guilty about it.

But what if you had to work until midnight last Thursday—a 16-hour shift!

Ok, do you do that every day? Or even every Thursday? I bet not. But if so, your work may be a necessary evil, and I covered that earlier.

I bet some emergency popped up and you pulled the double-shift to get things back on track. You can feel bad about that, but it's actually part of being in IT. There are unpredictable outages in IT. And that unpredictability applies to every profession. By definition, you cannot plan for the unpredictable.

It's ok to be disappointed when work intrudes into life. It's not ok to have that disappointment spiral into guilt. If you do, you then have a whole new problem to deal with. What's worse, it's a problem you manufactured.

I don't know about you, but I have enough to deal with already. I don't need to be building new issues to consume my heart, mind, and time.

Balance

Balance means that sometimes you lean this way and sometimes you lean that way. If you pulled a 16-hour shift last Thursday to bring a dead server back to life, you should be able to schedule an afternoon off this week or next.

Is that fair? No. Fair would be getting an entire day or day and a half off for the extra 8 hours you pulled.

Pay attention. This is important: Life isn't always fair.

Balance is important. So is recognizing you're working for a reason, and that reason is not selfish. There's no reason to feel guilty about that.

Balance and Transparency

I believe transparency is crucial to successful leadership. I share a lot from my life in my blog. I think that's a cool part of blogging, and I praise folks for being transparent. But I realize that everyone has a different tolerance for transparency. It's like a sweet tooth: Most people like sweets some of the time, some like sweets way too much, and some can't stand sweets. Similarly, most people share some things from their life, some folks share too much information, and others can't stand transparency. Most folks find themselves in the middle somewhere. I try to strike a balance with transparency online and as a team leader.

How do you know you're doing transparency right? Heck, I don't know! What works for me works for me, and may not work for you. If I'm getting requests for more information and also requests to share less, I feel like I'm in the middle of my target audience; that target being people who are comfortable with the amount of information I share.

I don't share everything. I don't even share everything technical. Some of the things I don't share are business related—things like personal projects I'm tinkering with that may have business value in the future. I don't share proprietary information at all. In addition, I don't share everything that's going on personally.

I draw my own line, and it's in a different place than the lines other people draw. My line leans to the transparent side. It's up to each of us to find our own balance between transparency and over-sharing.

Balanced Transparency and Leadership

Ask the folks on my team, and they'll tell you I'm just as transparent at work, too. Most of my team knows a lot about what's happening in my personal life. They know when I'm going to take a road trip and where I'm going. They know my kids—mainly because I work from home and my kids can (and do) occasionally, pop into my office when I'm in the middle of a conference call.

What does this transparency at work do for me? Well, it keeps life simple, for one thing. You get the same Andy in the community that you get at work. In the community, I Am Here To Help™. At work, I Am Here To Help™ too. (Note: Let me be transparent here: I wish I'd thought of the ™ symbols on my own, but I didn't. My boss started applying ™ to several phrases he repeats to his team often. I thought it was cool enough to steal the idea.)

I find transparency is contagious. It also breeds trust, respect, and loyalty. I've learned the best way to encourage loyalty is to demonstrate it. In other words, be loyal to folks if you want their loyalty. If you're transparent about your loyalty to others, you inspire loyalty back to you.

How do you pull it off? A few people know how to inspire loyalty, but this minority is far outweighed by those who know how to dead-end loyalty in its tracks. I find lots of opportunities to inspire loyalty by observing others and making sure to acknowledge and appreciate them.

An example: Once a project is complete and running in production, managers will be lining up to pat the development team on the back, buy them lunches, and tell them what a great job they did. Nothing is wrong with this, and it's perfectly normal.

The opportunity lies in the fact that while the work is in progress, these same managers are composing nasty emails and making threatening phone calls, and in general, not being very polite. What those managers do not realize is: The work they will later be patting the developer on the back for doing—the very developer they're currently threatening—is being done at the very moment the manager is behaving badly.

Don't think that the developer doesn't notice that discrepancy between the fact that the manager is behaving badly while the work is happening and then dishing out praise after the fact. That manager is missing an opportunity to be transparent and to engender loyalty during the development process.

That type of manager is simply punching the developer in the brain with this behavior. And make no mistake; this slows down development and consequently, delivery.

The opportunity for transparency and balance here is to praise the developers' midstream. While the project is still under development—and especially if things aren't looking good—let the team know—that you have the utmost confidence in their ability. The effect is astounding.

One important note: Don't lie. Don't tell the team you have confidence in them if you don't. They'll figure it out as soon as you lose your cool after you meet with the client and come yell at them. They're smart like that, those wily developers.

You can help your team deliver more and do it more effectively and sooner if you inspire them. Transparency helps, as does support during the winter months of a project's development cycle.

As with all things, balance is important with transparency. Know how much you're comfortable with sharing, and know how much your team is comfortable with you sharing.

Chapter 8. Workplace Hierarchy

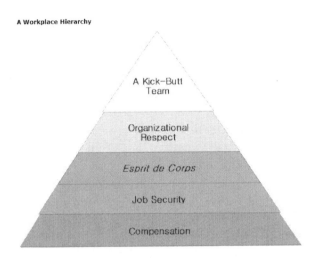

A Workplace Hierarchy

When I think about teams and hierarchy in the workplace, I borrow heavily from a Maslow-ian (is this a word?) approach. Maslow's Hierarchy of Human Needs has been applied (and misapplied) to many areas before. Someone has probably already made this connection for teams in the workplace. If so, cool! I haven't seen it. This my pass at applying it to teams.

The way the hierarchy works is that each tier in the hierarchy is dependent on the tier beneath it. The lower tiers on Maslow's Human Needs pyramid are basic necessities. The lowest tier is reserved for physiological needs. The second tier of the Human Needs hierarchy is safety.

In 1943, Maslow proposed that a human being is motivated by these needs from the lowest level to the highest. If your physiological needs are unmet, you will instinctively concentrate on those needs and ignore all else. If all physiological needs are met and you are unsafe (or feel unsafe), you will concentrate on becoming safe. And so it goes, up the Human Needs pyramid from physiological, through safety, love or belonging, esteem, and self-actualization.

I think of the five tiers of Maslow's Hierarchy of Human Needs as being similar to five tiers of the workplace for a team. In my version of Maslow's hierarchy

- The Physiological tier is analogous to Compensation
- Safety is analogous to Job Security
- Love or Belonging is analogous to Esprit de Corps
- Esteem is analogous to Organizational Respect
- Self-Actualization is analogous to building a Kick-Butt Team

Let's drill into each of these a little deeper.

Compensation

At the bottom of the employee pyramid is compensation. If team members do not perceive they're receiving adequate compensation, their priority will be to satisfy this need.

Some people will argue compensation is not at this basic level of need. They're right—to a point. If you have a crappy job, no amount (well, no reasonable amount) of money is really worth it. But money is like other things—oxygen, for example: It's not really a problem until you're not getting enough. People will strive to meet this basic workplace need (make enough money) until they make enough money to feel that they're being appropriately appreciated.

Some managers, directors, and executives believe some extra money in the form of an incentive bonus can make up for a deficiency on this most basic level of the hierarchy. But incentive plans can backfire with people who use exotic math (and by exotic, I mean long division). Consider the thoughts of an employee who just received a $250 performance bonus: "I get paid $42.50/hour for each hour in a 40-hour week. I did 100 hours above that and I got $2.50 for each of those 100 hours."

Mind you, $250 is something—and something beats nothing, especially when there's no 401k matching or raises. But was the need truly met?

Whenever I'm asked about compensation, I advocate keeping things simple. "Pay people what they're worth," I tell executives, "If you are going to provide bonuses, make sure they are truly *bonuses* and not considered part of the compensation package."

Job Security

Above money in the hierarchy is job security. This is more than, "Can the company survive without me?" If an employee is relying on that kind of job security, I have some bad news: The company *can* survive without them. It may cost the company money and time and the company may even rue the day they let an employee walk out the door, but the company will survive.

Other stuff plays into job security. For example, everyone wonders, "Does what I do matter?" People want to know if their work counts. More than that, as a manager, are you letting your team members know *how* their work matters? Job security in this hierarchy can be as much about an employee being secure in the knowledge that he's appreciated as it is about knowing that he'll have a job tomorrow.

Esprit de Corps

Next up on my version of the pyramid is belonging, or cohesion within the team. Everyone needs to feel accepted by their coworkers and by their managers.

The kinds of questions that will help you identify things that make team members feel accepted are: Do we share a sense of mission? Is there a vision? Goal? End date? Are we all rowing together?

This is where esprit de corps starts. The team hangs together, or the team members hang separately. A great discussion on this topic is the article, The No-Cost Way to Motivate, by Patrick Lencioni.

Esprit de corps is about a bond among team members. It's about their love for one another and their dedication to deliver an innovative and high-quality solution. It's about love.

How does a manager encourage love among the members of her team? A manager removes the barriers to love. What is a barrier to love? Actually, the correct question is, *who* is a barrier to love?

If a person on the team is blocking others, you have at least three possible responses:

1. Counsel them about their behavior
2. Reassign them
3. Fire them

To encourage esprit de corps, a manager can positively reinforce desired behavior. Desired behaviors include team members helping other team members by sharing part of the workload, coaching, and simply encouraging others. Teams self-organize, and these positive behaviors drive the outcome.

Since I have already bashed money as a motivator, how can a manager achieve positive reinforcement? Recognition works. It doesn't have to be public recognition. Take the emerging leaders of the team to lunch, or give them some time off, or sponsor some training for them. Employees love these rewards!

Organizational Respect

In my organizational hierarchy, Maslow's esteem tier reminds me of the respect of other teams. To gain respect from other teams, your team needs to communicate and demonstrate that they're worthy of respect.

Consider the following questions: Do other teams know what your team is doing and why? They know you're holding them up or badgering them for answers. Do they understand why? Are you being professional as a team in your interactions with other teams? Are you communicating as a unit?

A manager can greatly impact this level of the team dynamic. If you're managing a team of developers, uou can use project management methodologies, such as Scrum, to wrap organization around the team's communications to other teams. One nice feature about Scrum is stakeholders and members of other teams can be invited to attend. If the Scrum stand-up meetings are held to fifteen minutes, stakeholders and members of other teams will often attend regularly to catch up on the progress of a team.

A Kick-Butt Team

Take care of the basic needs of your team, and nature will take its course. You'll be stuck with a self-actualized team that you can throw at just about any problem and watch them solve it. And they will have grown organically. They'll be in bring-it mode. They'll crave challenges and hold friendly competitions to innovate on the best solutions. They will laugh a lot. They'll be characters.

There are several implications to using this hierarchy as a management tool. One is that if team members do not perceive they're receiving adequate compensation, their priority will be to satisfy this need by seeking employment elsewhere. Perception is reality, especially in this context. And it's the perception of the team member that counts, not the manager.

Be aware of where each member of you team fits in this hierarchy and work to make sure the needs at each person's level are being met. The key takeaway for managers from this chapter: Treat your team well, then watch them grow!

Chapter 9.   The 15-Minute Meeting

I'm a big fan of 15-minute meetings. The 15-minute meeting is all about multitasking, which I'm not good at. If I can't multitask, at least I can keep meetings short and focused, allowing me to parcel out my time in ways that let me get things done serially but efficiently. To accomplish this, I turn to my engineering background, which provides a strategy that computers use and that I can emulate: time slicing.

Multitasking

I have discovered something about myself: I stink at multitasking. It's true I do a lot of things, but the way they get done is not pretty. This is not a complaint; it is part confession and part recognition-of-reality. The recognition of reality is I have two hands and one mind. I can only dedicate them to a single task at a time. The confession is that I like it this way.

I'm a notorious monotasker. I remember when I discovered my inability to multitask in third grade. Have you seen the movie *For the Love of the Game*? While he's on the pitcher's mound, Kevin Costner's character focuses by saying, "Clear the mechanism." And the crowd noise fades to nothing. That happened to me at age 8 in my classroom in Burkeville, VA. We were given a page of arithmetic in math class. I started solving the problems, and everything else went away.

That oblivious immersion has happened since then when I'm coding or doing engineering work. I remember designing electrical control systems in 1990s when I ran a small industrial automation shop. I'd get to the office around 5:00 AM, make a pot of coffee, and start working. In an hour or so, I'd get hungry and eat something. Then I'd make another pot of coffee and work for another hour. And then I'd look up and it would be dark outside. I would think, "But I've only been at work a couple hours!" In fact, I'd put in 12 to 14 hours.

Time Slicing

Since I can only do one thing at a time (and since I really enjoy doing only one thing at a time), I need a solution to get more than one thing done in a given time period. My solution is an approach that computers take. It's called [time slicing](#).

Back in the day, computers had but one CPU. Can you believe that? A single CPU. It's true kids. I'm telling you. Look it up on the interwebs if you don't believe me. Back then, computers would do one thing for a while, and then go do something else for a while.

Computer processing has become more efficient since those days. Initial enhancements included the use of queues and schedulers to improve software performance, and faster clocks to improve hardware performance. Modern CPUs have multiple cores that allow true multitasking. But if we look beneath the core, individual circuits continue to execute a single instruction per cycle. Schedulers and queues are still there, executing quickly and thereby creating the illusion of multi-tasking while the computer is actually time-slicing.

Humans who are bad at multitasking can also mimic this behavior. Time slicing brings me to the beauty of the 15-minute meeting.

The Beauty of the 15-Minute Meeting

When I think in terms of time slicing, I think of how much time I can spend on a particular task and how many tasks I can accomplish in a day. 15-minute meetings break my day into 40 intervals (for a 10-hour day). If I go with 1-hour meetings, I only have 10 intervals. I can work on 4 times as much stuff if I slice my day into 15-minute intervals.

As a manager, I have to think about productivity and how people are spending their time. I find 1-hour meetings have 5 times as many people invited to them than 15-minute meetings. That's one of the reasons why the meeting is an hour long—to give everyone a chance to talk. And here's the interesting part: Most of any meeting consists of two people communicating while the others listen. For those of us who practice math, that means in a 1-hour meeting, a lot more people are listening, zoning out, or ignoring the meeting and doing meaningful, productive work instead of participating in the meeting.

15-minute meetings should never have more than ten people. And really, three or four is optimal. Identify the people who can best make the decision, get them together in a 15-minute meeting, and make the call.

If we have only 15-minutes to talk about something, we're not going to waste a lot of time. The very length of the meeting creates a sense of urgency.

What about the other people who need to know about the decision?

Send them an email.

Whether you're a multitasker or a monotasker, you can always use more productive time. And you want your team to use its time actually working. The solution? 15-minute meetings. They let you increase the signal-to-noise ratio in your day, and 15-minute meetings are a more effective use of everyone's time than hour-long meetings.

Chapter 10. Problems

I define a problem as an unsolved issue. It could be minor. Or it could threaten your project, livelihood, or even your life. Problems can be solved. Along the way, however, you have to be aware that people react to problems in ways that can complicate and derail efforts to solve the actual problem. Two destructive reactions are drama and terror. Once you know how to deal with these reactions to a problem, you can get down to fixing it. And fixing it begins with asking the right question.

The Effects of Problems

A problem can have many effects, but the worst effects are the ones that are manufactured by people. The most prevalent category of people-generated effects is drama.

My daughter Emma Grace has a cold. That's a problem. An effect of the problem of Emma's cold is that Emma is upset that she can't play in the snow with the boys. Now Emma could make a fuss about wanting to go outside and play despite her cold. She could create quite a drama in our household. That drama would not accomplish anything. Emma would still be sick and have to stay indoors. But the drama would exhaust everyone, including Emma. Instead of creating drama, Emma would be better off resting in bed and letting her body fight the cold.

Similarly, a business problem can create drama that wastes everyone's time and does nothing to address the actual problem. Say, a server crashes due to a hard drive failure. Your boss calls you screaming that it's your job to anticipate hardware failures and replace hard drives the week before they fail. The failed hard drive is the problem. Your boss's illogical emotional rant is drama.

Drama Is More Work

For everyone involved in the actual solution to a problem, drama is just more work. The time spent responding to the overreactions of others is really time that could be spent solving the problem.

Some Irony

When the problem is solved, the people manufacturing the drama will secretly congratulate themselves for their "contribution," believing the additional workload and stress they added to the hearts and minds of those actually solving the problem (punching them in the brain) helped in some way.

Nothing could be further from the truth. If you don't have your hands on or in the solution, you can do one of two things to help:

1. Get your hands on or in the solution.
2. Stay out of the way of those with their hands on or in the solution.

Overcoming Drama

How do you deal with drama when you're faced with it? That very question is flawed. You cannot deal with drama when you're faced with it: It's not within your realm of influence. It's nothing you can change. The root of drama is in someone else's mind, and you cannot affect that unless you're Matt Parkman.

Drama, when it's occurring, is best identified and then ignored. In the example above, explaining that the hard drive failure is the problem and the steps you're taking to replace and rebuild it should suffice. If it doesn't, you could be the victim of a crappy job. Personally, I don't even acknowledge the drama that attaches itself to problems.

Serious efforts at curbing drama can only happen before and after drama strikes. Before it happens, recognize and communicate the potential impedance drama adds to any real problem. After the fact, point out the problem. Be specific about the amount of time it took to solve the problem and the additional time wasted mitigating the drama.

Terror

The second people-generated effect of a problem is terror. This is when someone decides they need to stir up everyone in the organization when a problem occurs. You know the tactics of terror They include things like sending over-excited emails to everyone in the management chain, calling people in the middle of the night, and leaving threatening voicemails. Terror is another reaction to a problem that does nothing to solve it and wastes everyone's time. Let's consider terror in the context of a problem.

A Scenario

Your project is operating on the assumption that everything will go right. There are rules, and one rule clearly states that every task is assigned a primary resource, the person responsible for the project's success. A backup resource is also assigned. This is a person who, you assume, can perform the job function as efficiently and effectively as the primary resource.

Something unexpected and bad happens in the middle of the night. The primary resource is unavailable. The problem appears to be a new feature added in the past week. The backup resource is working the issue but not having much success because the backup resource has little knowledge of how the new code works.

The project management team realizes they stand at a roadblock. What do they do next?

False Assumptions

This scenario includes two false assumptions: The first false assumption is that everything will go right because you have accounted for all possibilities and have created rules to deal with contingencies. The second false assumption is that you will always have a backup for your primary resource.

The Unexpected Isn't Supposed to Happen

The first assumption is based on the idea that everything that can happen during a project cycle can be accounted for ahead of time and that nothing outside the defined assumptions and rules can occur. What happens if, for example, six feet of snow falls in two weeks where one foot of snow is the norm? You can't account for the weather in projects, but you can leave time in the schedule for the unexpected.

Even the best laid schemes of mice and men often go askew. I apply the matrix below to all plans. The idea is to remind me that stuff happens. Sometimes it's good stuff and sometimes bad, but it happens nonetheless.

I use the table below to think about the stuff that happens. The upper left quadrant represents the driver for the plan. The upper right quadrant is also part of the plan. It represents the stuff you expect to sacrifice to achieve the planned goal.

	Positive	Negative
Intended	Intended Positive Consequences	Intended Negative Consequences
Unintended	Unintended Positive Consequences	Unintended Negative Consequences

The lower left quadrant is awesome and usually results from a miracle, good karma, a lucky break, or a well-designed process. The lower right quadrant can sink your project, cause untold misery, or even kill you.

How do I use this matrix to plan projects? Experience. For example, from experience I know projects rarely have *positive* unintended consequences, especially when compared to negative unintended consequences. In my experience, the ratio is roughly 8:1 in favor of the negative. If a software development project is using an Agile software development methodology (such as Scrum), I add 20 percent to an everything-will-go-right project plan. If a project is not using an Agile methodology (even if they *say* they are using an Agile methodology), I break out the serious multipliers: 2x, 3x, 4x, etc.

Why does software methodology make such an impact? Agile methodologies empower the developer by granting them the authority to move deliverables into and out of scope for a given iteration (or Sprint, in Scrum-speak). Test-Driven Development (TDD) is often employed by Agile developers to identify and isolate bugs *during* the development cycle. Other methodologies rely on testers to perform similar tests, report the bugs, and then have the developers address the bugs. The difference in my methodology multipliers is the time spent sending bug-y code to testers, waiting for them to test it and write up the bug report, and then waiting (again) for the developers to respond.

This approach also applies to projects that are not in the field of software development. You know the factors in your area that will cause the multipliers to grow ever larger. It's all about agility in the regular sense of the word.

Every Primary Resource Has a Backup Resource

The assumption is that you'll always have a backup person as a way to prevent single-threading and single points-of-failure. Is this realistic? Yes, provided you calculate the number of people you need for the project and then promptly double that number.

If you practice pair programming, then and only then does everyone have a backup as competent as the primary. If your project demands a backup for each primary without practicing pair programming, you can achieve the effect of doubling resource intelligence by building time for code review, documentation, and similar efforts into the project deliverable schedule.

Software projects rarely take time for these measures. Why? It's a communications issue. When a project manager asks a developer, "How long will it take to develop feature xyz," the developer hears, "How long will it take to develop feature xyz?"

I know. Crazy.

What the developer doesn't realize is that the project manager is really asking, "How long will it take to dissect the customer deliverable expectations; provide feedback to the business analysts until you and they agree on the deliverables; develop the software; perform unit, integration, functional, and performance tests; package the software for deployment; and complete the documentation for feature xyz?"

It's all in how each participant in the conversation defines the word "develop." The point is that you need to account for false assumptions and communications failures when you plan your project. If you don't, you're laying the groundwork for terror.

When Terror Sets In

Let's go back to the assumption I opened this chapter with: The project is staring at a roadblock, on one hand, and you have a non-responsive single point-of-failure on the other. Actually, the "project" isn't doing anything. People make up the project team. When you tell them they're blocked and there's no backup, that's the start of the terror cycle.

Someone doesn't know what to do or doesn't want to be held responsible, so that person passes the fear up the management chain and throughout the organization. The frantic project manager begins the escalation ritual. A call in the middle of the night, a voicemail message, followed by an email message, followed by subsequent calls. A manager is called. If that manager cannot be reached, that person's manager is called.

Terror Makes More Work

Like drama, terror really isn't worth the trouble. Common sense dictates the number of phone calls, nasty voice mails, or threatening emails does nothing to move the problem toward resolution. In fact, punching the developer in the brain slows down progress because the developer now has more stuff in her head than solving the software issue.

Some Irony

Again, as with drama, when the roadblock is overcome, the people manufacturing the terror will secretly congratulate themselves for their "contribution." They will convince themselves that the additional workload and stress they added to the hearts and minds of those actually solving the problem helped in some way. Nothing could be further from the truth.

Overcoming Terror

How do you deal with terror when faced with it? The best solution I've found is to show up and solve the problem, ignoring the emails and deleting the voice mails. Be professional. Be polite. Rise above the terror. Do what you know is the right thing to do, and do not do anything you know is the wrong thing to do.

People are creatures of habit. Terror is a habit. It can be broken if you demonstrate that it simply doesn't work on you. If you calmly go about your job, doing the very best job you can—and not giving in to the temptation of terror—a funny thing happens. People start respecting you and treating you better.

Will you always have the answer that calms the waves and silences the storm? Will you always be the hero? Will you always show up with the silver bullet? Probably not. But if you refuse to react to the chaos some will hurl your way, you will communicate more with your actions and your words.

The Right Question

Drama and terror are not constructive ways to respond to a problem. The proper response to a problem is the right question. In contrast to an emotional rant and the resulting drama, when a hard drive failure takes down a production server, the right question can move you towards the immediate solution, more preparedness, higher availability, and less downtime.

If you can minimize the drama—or circumvent it entirely—your team can concentrate on solving the problem and then improving the process. This can start a positive cycle where one good deed inspires and fuels another, and that another, and so on.

Often, identifying the problem is immediately preceded by asking the right question. Why does this help? Framing the question forces your mind to order the facts. You have to consider what you know and what you don't know to ask a good question. In fact, the right question will take into account all you know and ask about all you do not know.

The right question may not be exhaustive. It could be about the next step in a series of unknown steps. Does the right question have to include all the other unknown steps? Nope. If you know there are a host of unknown steps, the right question implies moving closer to the other steps in the solution by identifying the nature of the next step.

When you find a problem, throw a party. Why? Finding problems is the hard part. Fixing problems is easy.

Someone recently asked me how long it would take to fix a software issue. My response was, "Fifteen minutes. It might take me a week to find it," I said, "but I'll be able to fix it in fifteen minutes."

Software is like that. Welcome to my world.

Cutting through the symptoms and excavating effects until you reach a root cause is a lot of work. There are a million subtleties to examine and rule out. And to make matters worse, problems of similar complexity can take vastly different amounts of time and resources to locate.

Why? Excellent question. My answer: I don't know. The best I can do is warn you of the observed trend. I often use an analogy to communicate this: It's like looking for your car keys. They're always in the last place you look.

Some astute individuals point out that when you find your car keys you stop looking. To this, I reply, "Exactly."

Process Improvement

Once you and your team are focused on asking the right questions and preventing drama and terror, you find that asking the right question can go beyond solving a specific problem and lead to process improvement. The right question can bring out overarching problems with process that contribute to the specific problem you're trying to solve.

I remember my first stint as a manager. I met with the team and asked, "What do we do?"

Most of the responses were along the lines of, "All day long I put out fires."

I then asked, "What do we want to do?"

Again, most of responses shared a theme: "Be proactive, not reactive."

One important note here: The team already knew what they wanted and needed to do. They didn't need a manager telling them, but I was asking the right questions.

Process improvement is all about going from reactive to proactive. The first team I managed pulled this off by asking the right question, which was, "How do we get from reactive to proactive?"

We identified a metric, DBA-hours, and started tracking our activity. Applying some analysis, we found 90% of our time was spent reacting (e.g., making changes to data) and 10% was spent on developing long-term solutions (e.g., altering a stored procedure to properly insert the data in the first place).

We identified a crucial middle-step. We called it "fighting fires better." To facilitate this, we documented procedures for common, recurring fixes. We created a script library that started with hardcoded scripts and grew into parameterized scripts. This work later served as the basis for most of the solution development. This illustrates how you can go from asking the right question to improving process and then to solving specific problems.

I view problems as challenges. When I'm feeling clever, I refer to obstacles as "something to overcome." By recognizing that problems have effects and that some effects, such as drama and terror, are generated by people, I can take a step back and make sure my team understands the difference between a problem and its effects. I can help the team ask the right questions and save time, effort, and frustration.

Chapter 11. Software Is Organic

Can software be manufactured? Pull up a chair here beside Grandpa Andy, and let me tell you a little story.

A long time ago, back when the years began with the number 1 and the US had a manufacturing economy, quality became important in manufacturing. Terms such as "Six Sigma" and "ISO 9000" became common in manufacturing. In the US, manufacturing plants bought into a concept called "quality" in a big way. There were classes, seminars, studies, and books—lots and lots of books—about applying the principles of quality in every industry.

I was working in manufacturing during this frenzy. I took the classes, attended the seminars, participated in the studies, and read a lot of the books. I discovered that, embedded in Six Sigma and other quality programs and procedures were ways to predict stuff like the number of widgets that would be produced and their quality. This was called predictive analytics.

I have a confession: I have always enjoyed predictive analytics. For an engineer who enjoyed predictive analytics, all this quality stuff was awesome! I really enjoyed using quality processes to improve the performance of manufacturing processes.

And then, manufacturers thought, "See these results? This is fantastic. Let's do this quality thing everywhere!" So they implemented process metrics collection in maintenance departments and inventory management. And one day, quality was implemented in IT.

A funny thing happened in IT: Quality programs didn't work. Why? IT lives! Software is organic. It starts small—germinates as an idea really—and then matures. It grows.

How do you measure software's progress? How do you measure anything organic—say, a tomato plant? If you're measuring a tomato plant, you can measure its height and record that each day. Is this a valuable metric? Probably not. What you really want to know is:

- When will it produce fruit?
- How much fruit will it produce?
- How long will it produce fruit?

Will measuring the rate of growth tell you any of this? Not really. Measuring the plant height will tell you it's getting closer to producing tomatoes, so that's something. But since tomato plants of varying size produce fruit, this metric is lacking.

Back to software: What were the quality metrics for software development in manufacturing IT departments?

- Lines of code
- Bug count
- Service Level Agreement (SLA), or deliverable schedule met

As with measuring the tomato plant's height, these metrics may tell you something. But is what they tell you meaningful? Let's dive into this further.

One way to develop software is to break it into three phases:

1. Frame
2. Build
3. Optimize

Counting lines of code could work in the Frame and Build phases. Lines of code could indicate productivity in these phases.

But in Optimize? A really good optimization may involve removing lines of code. So this metric backfires in a major way at a crucial time in the development lifecycle.

Bug counts may sound like a meaningful measure. You want your software to work without any problems, so number of bugs seems like an indication of quality. But bugs are not linear and therefore not a consistent metric for tracking quality.

Hitting an SLA, or deliverable deadline? This is a good metric, but it's only available at the end of the project so it doesn't help you use quality metrics to increase productivity or progress.

So if traditional quality metrics don't work for software, what do you do if you're sold on measuring quality?

Listen to the Developers

Software evolves throughout a project. (It's alive, remember?) Software cannot be manufactured using assembly-line approaches—at least not in 2014. This is why manufacturing experts who were also experts at quality could not make the same quality metrics work in their own IT departments.

The evolving nature of software is one reason Agile software development methodologies work: Agile accounts for the true, living nature of software development. And that nature does not lend itself to repeatable and deterministic measurement.

Who are the best judges of software quality? The developers. Listen to them.

For example, consider evidence-based approaches. These are a neat blend of metrics collection and paying attention to the developers. Consider a Work Breakdown Structure (WBS) approach to software (which also works really well in other engineering disciplines, by the way). WBS is forward-looking, attempting to extrapolate the effort and duration required to build components of a software project.

WBS is not so much extrapolation as it is addition. Break down the tasks into manageable chunks, estimate each chunk, and then add up the numbers. Simple right? Too simple, in fact. But I'll get into that later in a bit.

Battle of the Three Letter Acronyms (TLAs): WBS vs. EBM

Contrast WBS with Evidence-Based Management (EBM). Like WBS, EBM breaks software development into components and phases. Unlike WBS, the evidence-based approach relies on historical data gleaned from the first (or first few) development efforts. Effective metrics can be collected early in the development phase by allowing developers to do what they do best—develop software—while measuring how much effort and time the development effort takes.

If you want to get all crazy, you can ask the developer for estimates on effort and duration before she starts. Then, when that chunk of code is built, a good project manager can compare the estimates to the measurements.

It's vital not to sabotage this cool start by holding the developer's feet to the fire. The phrase "But you said it would only take ___," should never cross your mind, much less be communicated to the developer. You're trying to build a performance characteristic here. What you want is a way to predict (as in predictive analytics) when the project will complete.

Good project managers make this claim all the time: "I just want to know when it will be done. I don't care how long it takes. I need a number I can communicate with confidence to the client."

If that's true, here's your chance to prove it. Collect the developer's estimate and then collect the evidence (the actual effort and duration ex post facto), and then do the math. Developers are creatures of habit. Some will guess high; most will guess low; every now and then someone will nail it. It's a guess.

If you pop enough of these metrics into statistics functions on your favorite calculator, you will see a pattern emerge. And this pattern can be an estimated: performance ratio you can begin to apply to each developer's estimates. (Like I said, crazy...)

The same approach applies to software teams and tomato plants. With software teams, the more I observe, the more I understand about how the developers work. In time, I'll be able to predict the progress of the software development process.

With tomatoes, I've already done a lot of observation and can draw some conclusions. For example, I know that if we wait a couple weeks beyond the recommended planting date for tomato plants in Farmville Virginia, we get more and better tomatoes out of our garden plot. Why? It could be any number of reasons really: The amount of sunlight that hits our garden, the soil temperature and composition, how we manage weeds and pests. I'm certain all that plays into the result, but the result is what I'm after—and that's more and better tomatoes. I don't need to be an agriculture major to grow tomatoes. I just have to experiment, measure, and record the results.

The tomato plants, like the software developers, know what they need to do. Our job is to make them as effective as possible. That's where management comes in.

Organic Software Management

The beauty of organic software management is that the software and teams perform as they should. Can you force a management style onto a team? Yes. Does it produce? Sometimes, but rarely as well as organic management. Why? Because people self-organize. And they do so instinctively, organically even.

Geeks are people too. Whenever I think about how geeks socialize, I am reminded of a quote from the movie *I, Robot* (not sure if it was in the book, but I find the robot series from Asimov very inspiring): "Why is it that when robots are stored in an empty space, they will group together, rather than stand alone?"

My family raises chickens on our small farm in Farmville Virginia. (I am not making up the name of the town, I promise.) These are egg birds, not for meat. We also have a small garden. We derive immense satisfaction from growing our own food.

It's amazing how versatile nature really is. If you get a few key things right, you can do almost anything you want to a garden and still get vegetables and fruit. If you don't do those few key things right—and all of them, mind you—you will get little if anything from your garden.

The first of those key things is creating the right conditions. The second key is planting the right stuff at the right time. The third key is taking care of the garden. Our garden is my model for using the word "organic" as applied to software development.

So how does organic software management apply to teams? Ask yourself the following questions:

1. Are the conditions conducive to producing the desired outcome?
2. Are you introducing the right stuff at the right time?
3. Are you taking care of the team / project?

Conducive Conditions

When you think of the conditions required to produce quality software, what do you think of? I think of a comfortable and quiet workspace, modern equipment, and reasonable management.

Some folks can code away in a noisy, busy environment, and some even have to. Others cannot. I'm one who needs quiet. If I'm not in a quiet environment, I create a controlled noisy environment through the magic of my media player. Be sure you know what kind of environment your people need, and do everything you can to provide it.

Nothing breaks my heart more than seeing developers tasked with creating tomorrow's applications on yesterday's equipment and software development platforms. The people who hired those developers must think they're super-awesome-coders. I bet they're paying them as if they're super-awesome. Probably not. Probably someone is saving money.

My favorite anecdote about saving money involves toilet paper: It keeps the crowd on the edge of their seats.

I've worked with computers and development tools that are too slow. The response to my click took so much time I sometimes forgot why I clicked in the first place! Substandard equipment does not save money, it costs time.

Right Stuff, Right Time

When it comes to managing a team of developers, you need lots of communication to make sure you're aware of what's going on. For example, good people will end up being asked to do more work than others. This will start a cycle that ends with intervention from management or from the good developer. Usually the good developer manages being overloaded by quitting, so it behooves you, the manager, to get involved before this happens. You need to make sure that good developer is keeping you informed about workload concerns.

Good developers need breaks. Most (but not all) good developers love learning new stuff. Training is a win-win for these folks. They get a break, and learn something new that can help the team and organization. Make sure you're communicating that training opportunities are available, and make sure you know who needs a break and when.

Also, developer teams require steering. It's not suicide-knob steering here; think tiller on a large boat. Gentle guidance, not wild thrashing. And definitely not too much. Over-steering can get you into trouble faster than anything else, especially if you're on a slippery surface. Again, communication is crucial to knowing how well you are steering.

Taking Care of the Team

In my family garden, we do things to keep out stuff we don't want. We weed, we apply as little pesticide as we can (erring on the side of losing plants rather than over-treating), and we shoot at the neighbors we catch stealing stuff. Ok, I made that last part up. Our neighbors are welcome to anything they want from our garden. We only shoot at people we don't know (or readily recognize), and even then we fire a warning shot first.

Keeping out the wrong stuff is also important in managing a team. Everyone is not a good fit for every team. Taking care of the team sometimes means letting good people leave simply because they're not a good fit for the current operation. It's like weeding. Other times people outgrow the team and there's no place inside the organization for them. Occasionally folks blossom and take root in other sections of the organization. That's transplanting.

I see these patterns all over organizations. The organic approach is respectful of everyone on the team. The team doesn't have to be alone, but can stand together in an organic way, like Asimov's robots.

Chapter 12. I Don't Work on My Car But I Still Work on My Truck

I have a 1990 Jeep Comanche with the Pioneer package. My wife Christy kissed me for the first time in that truck. I love that truck. I don't do major surgery on my pickup, but I change the oil. I tune it up—plugs, wires, distributor, air filter. Christy owned a 1998 Accord when I met her. She now drives an Odyssey. I do very little work on the Accord, and I don't touch the Odyssey. Why? Because things change, and I wouldn't know where to begin working on the Odyssey. My skill as a mechanic is not up to today's state of the technology.

Things Change

I feel the same way about change and software as I feel about cars: If your experience with software isn't up to the latest level of technology, you don't have the knowledge you need for today's software. In fact, software changes even faster than cars. I've often said one software year equals ten car years.

I started flipping toggle switches that represented bits just about 35 years ago. Does 35 years' experience building software really mean that I'm automatically qualified to work on today's software projects?

Yes and No

Yes, I have a large and deep bag of coding tricks. It's difficult for me to quantify how much I call on that experience, especially the older knowledge. More than the knowledge itself, the decades of experience give me confidence I can architect software that works well.

And no, I do not think I call on the older knowledge often enough for it to make a major difference in my developer skills. In fact, I rarely call on any old tricks that I learned more than ten years in the past—and probably five years is the practical limit. Things change too fast for most of those old skills to be useful today.

Because I know that software development skills get stale quickly, when I interview developers I look at what they've done in the past eighteen months. That's about how long it takes to become a proficient software developer, and still being up to speed on the latest techniques. Two and a half to three years developing enterprise-class projects, or four years maintaining similar-sized applications, is the amount of experience I expect for a senior developer, and I want that experience to be fresh.

"I Used To Be a Developer"

When I hear, "I used to be a developer," on a project, someone is usually trying to tell me they know how to do a developer's job. They are saying, "I can do that job. I used to do it."

But that's not accurate if that person has been out of the developer field for more than eighteen months. Software changes that much with each release, and new design patterns emerge in the field every six months or so. Unless the person telling me he or she has recently been coding on the side, I'm not buying it. Sorry. Not really.

If you want respect from developers, here's a suggestion: Demonstrate respect. Think about how you would feel if a developer said, "Your job isn't that difficult. I bet I could do it."

Odder Still

Here's the odd part: I don't hear this arrogance from folks who stopped developing software a year or two ago. I hear it from folks whose last line of code was COBOL. I have nothing against COBOL. But batch programming from decades past bears little resemblance to multithreaded, Object Oriented, stateless, or even RESTful applications of today. (Note: I did not write, "no resemblance.") Today we have a different coding world.

One advantage of coding over the past 35 years is perspective. These days I don't call myself an application or web developer; I won't insult the field by claiming to be part of it. But do I create applications? Yep. Services, too. But coding up a single-threaded GUI does not a developer make today. I recognize my limitations.

I can work on my old Jeep pickup but I do not work on the Odyssey. If you used to be a developer and changed jobs, you are now something else. I'm sure your experience informs your current occupation. But do not confuse experience for expertise, especially in the rapidly-changing field of software development.

Chapter 13. A Turning Point

This chapter is about converting a negative spiral I your personal and professional life into a positive spiral. I'm not big on modern trends, and I don't like change. I don't think I need to tell you that. Just take a look at my hair (well, what's left anyway) and the way I dress. But I see some benefit in changing the stuff I can change. Namely, my attitude.

Years ago I found this blurb by Chuck Swindoll inspirational:

> "The longer I live, the more I realize the impact of attitude on life. It is more important than the past, than education, than money, than circumstances, than failures, than successes, than what other people think or say or do.
>
> "It is more important than appearance, giftedness or skill. It will make or break a company ... a church ... a home.
>
> "The remarkable thing is we have a choice every day regarding the attitude we will embrace for that day. We cannot change our past ... we cannot change the fact that people will act in a certain way. We cannot change the inevitable.
>
> "The only thing we can do is play on the one string we have, and that is our attitude ... I am convinced that life is 10% what happens to me and 90% how I react to it. And so it is with you ... we are in charge of our Attitude."

Secrets to Happiness

If you look around at work and life in general, plenty of things can discourage you. Swindoll's quote says to me, "You have a choice about how you react." Personally, I've made a conscious decision about how I will react. It boils down to: Don't do misery.

I did some misery in the past, and I think that's enough for one lifetime. From here on out, no more. When life hands me lemons, I give them to my lovely bride Christy, and she makes a tasty lemony dessert out of them.

Another conscious decision I've made: Do not let people live rent-free in your head. Sorry, them's the rules. I'm giving notice to anyone trying to live in my head: If you're in there, you either need to pay up or move out.

A Large Shovel

Although I've done some misery, I've also had some success in life and career. Someone recently asked for my philosophy on getting along at work. My answer comes back to attitude, and it's pretty simple: When I start a new gig, I look for the biggest problem in the department. I search for the dirty work. My attitude is clear. I'm willing to do whatever it takes.

Most of the time, I find helpful associates at the new gig who are more than willing to dump the garbage heap on the new guy—especially since he's a fat guy with facial hair who sounds like Kermit the Frog would sound if Kermit were from the South.

What do I do once I've found or been assigned to this pile of dirty work? I start shoveling. I dig in and get dirty. I ask questions and write down the answers. I think about the problem. I ask others for their opinions and insights. I learn.

And then I solve the problem.

Do you want to make a good first impression at a job? Solving a tough problem will accomplish that. You don't have to be the newbie for this to work. You can simply start being a new you. You can do it tomorrow. Change your attitude. Watch what happens.

Does This Work?

If you want proof that this approach works, here are some metrics. In the past decade, I've been hired a couple times by large companies. I've been promoted to manager both times. One time it took 93 days. The other promotion took less time. So yes, picking up a shovel and digging into the largest pile you can identify works.

All the Time?

I am saddened to report to you that I do not always do this. I occasionally get overwhelmed, lose my cool, and drop my shovel. I catch myself getting miserable and letting others live rent-free in my head.

As soon as I catch myself doing this, I stop. Nothing frees the mind like a 15-minute walk outside. I take a break and think about something else. And before I re-engage, I remind myself that life isn't about the accumulated quantity of times that Andy falls down. Life is a running sum of times that Andy gets up plus times Andy falls down, and that that total has just shifted back to its appropriate value: +1.

Maybe you don't want to be a manager. That's fine. But you do want to do a good job, don't you? These days bonuses and raises are not a given, but how about the value of someone recognizing your hard work?

The best part is elementary: Others can take your job, your position, and your stuff. No one can take the knowledge you gain. Your attitude is the key. Changing your attitude is the way to start a positive spiral in your relationships, life, and career.

Chapter 14. Now if You'll Excuse me, I'm Going to Go Screw Something Up

As I said in the previous chapter, I don't like change. But whether you like change or not, it happens. The key is whether you learn from change—and failure.

You may not realize it, but something has already changed. It happened while you were sleeping last night or the night before. It's going to impact your life in the next decade in a way you cannot specify now —or maybe even comprehend. What exactly happened? I'm not sure, but I know it happened. How do I know? Because change happens all the time. The world continues to flatten. Ideas emerge. Human interaction evolves. Computer-assisted communication improves. Social Intelligence blooms. Waves form.

Does this change scare you? Or does it excite you? You will react or respond to this change. You will make a choice. Your choice may be to simply ignore the change. In the words of Neil Peart (sung by Geddy Lee): "If you choose not to decide, you still have made a choice."

What will your choice be? What's the right choice for you? Those are questions only you can answer, and your answer will come from who you are.

Crap Happens All the Time

Some changes manifest as failure. You can decide whether failure is a good thing or not, but I've learned that failure is good. Let me explain that.

The other day, our team was having fun with an email thread about how some other consulting companies operate and how they treat employees. Our team was outraged at some of the things other companies subject their consultants to. Then, Tim Mitchell, one of our team members responded to the thread with the following:

> "We have a culture of failure.
>
> "How's that for a headline?
>
> "I read this article this evening, and loved it: http://www.itworldcanada.com/blog/why-cios-need-to-learn-to-use-the-f-word/95732
>
> "I remember working in environments in the past where everyone was terrified of screwing up, and therefore, nobody took any risks. Even worse, when the inevitable failure did occur in those environments, it was more about who to blame than actually finding and addressing the root cause.
>
> "I believe we have a culture of failure. More specifically, I believe we have a place in which we are free to fail at things that may or may not work, without feeling the weight of the world come down on us.
>
> "A funny thing happens, though, in a culture where you are free to fail: It leads to more frequent successes. When we are free to say, 'Yeah, I messed that one up. I see what I could have done better. I'll do better next time,' we learn. We truly want to do better next time, not out of fear of punishment or belittlement but out of a desire to improve ourselves and our team (who also fail and learn).
>
> "Andy wrote something the other day similar to the message in the article above. If we're not failing, we're not learning. I'm happy to be in a place where I'm free to fail, along with a lot of other people who are, as well.
>
> "Now if you'll excuse me, I'm going to go screw something up."

You can't get better praise for the power of failure. But when you're not doing much learning or the learning isn't risking next month's mortgage, it's easy to sing along with the chorus that goes, "We learn from our mistakes."

What happens when *you* risk and ~~fail~~ learn? Does the tune change? For some, it does.

Many are in Crap-Avoidance-Mode (CAM). A simple test for CAM is a little thing I like to call a "problem."

When a problem arises, people in CAM get upset. Why? They did not expect a problem to get through their safety net of crap-avoidance. People in CAM tend to create problems like drama and terror. Whether they react or respond, they start from the wrong mindset: "This wasn't supposed to happen." It doesn't matter whether it was supposed to happen. It did happen.

A few live in Crap-Management-Mode (CMM). Folks in this category actually *expect* crap to happen and aren't taken by surprise when it does. Quite the opposite: When crap isn't happening, they're enjoying the temporary vacation from crap, knowing it will return after this respite.

I'm not talking about optimists and pessimists. I'm talking about realistic expectations. I am battling the universe if I believe crap won't happen to me. The happiest people I know operate in CMM; the saddest, in CAM. I didn't write the Second Law of Thermodynamics, but it durn sure applies to me. Entropy isn't just a good idea, it's the law.

This Is About the Future

Things are going to change. Some idea that formed while you were sleeping is going to change your life. As the cost of implementing ideas drops to zero, change is going to happen more and more. Things are not only going to change, the speed and effect of the change is going to change.

Is this good? Maybe. I'll have to sleep on it (while someone is changing my future).

Enter Experience

My personal philosophy on change and failure has been shaped by my experiences. That's for sure. I'm in CMM almost all of the time. I expect crap to happen and accept it's my job to deal with it when it does.

My lovely bride Christy has a wonderful take on experience and judgment: "Good judgment comes from experience, and experience comes from bad judgment."

You have to roll with change. Paraphrasing an interesting quote I read years ago: "Surfers ride the waves" (from near the beginning of Rick Warren's book The Purpose-Driven Church). Surfers understand they do not make the waves. Waves come and go. They're as natural as anything.

Waves are a wonderful metaphor for change. A wave started last night while you were sleeping. You can complain about all these blasted waves. Or you can learn to surf.

Chapter 15. Getting It Right the First Time

This chapter is about getting software right by treating people right. Everyone wants to produce the highest-quality software. How hard could it be to meet the top industry standards? After all, hardware can be configured to perform with high availability and reach five 9s. Why can't software work without flaws?

A group of developers out there does just that: The onboard shuttle group of Lockheed Martin writes nearly flawless code: "The last three versions of the program—each 420,000 lines long—had just one error each. The last 11 versions of this software had a total of 17 errors. Commercial programs of equivalent complexity would have 5,000 errors."

(And it only costs $35,000,000 per year.) You can read about it in this article: "They Write the Right Stuff".

But the point is that the onboard shuttle group is not driven by relentless task masters who force them to work long hours. No, this group focuses on a process that enables them to produce almost flawless code without sacrificing their personal lives:

> "It's the process that allows them to live normal lives, to set deadlines they actually meet, to stay on budget, to deliver software that does exactly what it promises. It's the process that defines what these coders in the flat plains of southeast suburban Houston know that everyone else in the software world is still groping for. It's the process that offers a template for any creative enterprise that's looking for a method to produce consistent - and consistently improving -- quality."

I found the group's working hours striking: "It's strictly an 8-to-5 kind of place—there are late nights, but they're the exception."

What? They're writing nearly perfect code *and* working normal hours? I hear you thinking "How can this be?"

Working lots of hours does not mean better code. It may mean *more* code. And it certainly means more hours billed to the project. (Pay attention if your bonus is tied to return on investment—ROI.) But it may not mean *better* code. In fact, there's a point of diminishing returns when more people developing and more hours of coding begin to hinder the success of a project, pushing the goal farther from achievable and lengthening the time to delivery. The ROI begins to decrease.

Process

You have to always remember software development is an intellectual pursuit. We're talking about knowledge workers. The knowledge workers in the onboard shuttle group have addressed the software quality problem by focusing on process.

If you're reading this and you manually load pulpwood trucks all day, I know this will sound silly. After all, how hard is sitting down in an air-conditioned office typing and moving a mouse? I hear you.

Manual labor is work. I call it "real work" because I grew up on a small farm. Clearing land meant hooking a chain around a stump to pull it out of the ground with the tractor. Gardening wasn't fun. If we didn't weed the potatoes, we had less food for the winter. Hunting was something we did for food. If we wanted spending money, we pulled tobacco for $1 an hour.

But I bet anyone who does manual labor also has a process that lets them get the work done as efficiently as possible and with the smallest amount of errors possible. Everyone has their way of doing things, and that "way of doing things" is their process.

Intellectual work isn't like manual labor with respect to physical demands. But you can still demand too much of an intellectual worker. Just like you wouldn't ask a laborer to pull tobacco for 24 hours straight, you wouldn't ask a DBA to work 24 hours straight, either. (Or if you do, it should be the exception and not the rule.) However, both a knowledge worker and a manual worker can benefit from processes that streamline the work and help prevent mistakes.

The alternative for software development is to ignore this advice and burn-out good people; produce lower quality code that takes longer to test, troubleshoot, and repair (by already exhausted teams); and lose revenue.

Bottom-line: You cannot deliver good code on a death march. You can do other things: run off good people and deliver junk on time. But you cannot deliver good code. Think about where your focus is when you're managing a team. If you're focused on pushing to achieve more and more, maybe you should think about how a better process could make your team more productive and effective and allow them to work fewer hours. Use the work of the onboard shuttle group as an example.

Chapter 16. One-Time Productivity Boosts

Lockheed Martin's onboard shuttle group has mastered both productivity and quality by focusing on process. However, that group's success is not the norm. You can learn from the attempts that have been made in the past to find the magic formula for productivity. This chapter is about attempts to repeat tactics for increasing productivity.

Between 1924 and 1932 experiments about human productivity were conducted at the Western Electric Hawthorne Works near Chicago. One set of experiments involved illumination and serves as the basis of what became known as the Hawthorne Effect. The experimenters tested the premise that more light would increase worker productivity. The results of the experiment supported this hypothesis: Worker productivity increased measurably when more lighting was added.

But then the experimenters decided to do something different. The lighting was returned to normal after the experiment concluded, and the observers continued measuring productivity. A funny thing happened:
Worker productivity increased again.

The Cause of the Effect

There are books about why productivity increased under both conditions, and there are as many opinions as opinion-holders. I fall into the camp that attributes the effect to a combination of two things:

> 1. Change: Anything that increases attention to employees communicates, "We're interested"
> 2. The placebo effect: The workers knew they were being measured, so they thought something must be in place that would affect their productivity. And the attention inspired or motivated them to do more or better work.

Do you see the common thread between these two explanations? Attention!

Why is attention important? Attention communicates care and concern. It conveys importance. If the boss is worried about something or wants more (or more frequent) feedback, the line worker gets it: "This is hot and requires focus."

I think the increase in productivity resulted from a change in demonstrated interest (increased attention), and not from the mechanism used to deliver the message (increased or decreased illumination).

Does this effect of paying attention translate to knowledge work? Yes, but how do you sustain changing levels of interest?

Turn It Up

One way to let knowledge workers know you're paying attention is to constantly increase the pressure on a project. This is easy if you have a deadline, but it is not sustainable for two reasons: First, the deadline will pass (hopefully with the desired work or project completed). Second, cranking up the heat (the amount of attention) yields asymptotic results (i.e., the turnip only has a finite amount of blood).

The Value of Change

Are there other ways to communicate attention without turning up the pressure? Yep. You can alter the work environment in ways that don't punch developers in the brain.

When you're working with developers, one way to accomplish this is to introduce pair-programming for some period of time. Or, in just about any environment, you can alter the location where work is performed. Both changes will yield results the first time you try them. Diminishing returns will kick in, however, and the results of such tactics will prove as asymptotic as any other over time.

In other words, boosts in productivity decrease with each reuse of a technique. This eventually reduces the effectiveness of change as it's tried again and again over time. It's why I refer to these as one-time boosts.

The Central Rule for Projects™

It's important to remember the Central Rule of Andy's Projects™: The energy pumped into a project is conserved, and people are the only medium that this energy rests in. This means the pressure you apply to people working on a project remains with the project until the project is complete. You will not be able to sustain any increase in productivity derived from applying pressure. You can find new ways to increase the energy, but each new way will eventually become less effective.

To counter the effects of this rule, I believe in finding positive ways to demonstrate you're paying attention as a manager. Even though you know each new demonstration of attention will become less effective over time, you're building good will and trust. These are the things that will improve productivity in the long run.

Chapter 17. Institutionalized: A Geek Tragedy

This chapter is about the decline of an innovative organization. Good intentions. The road to hell is paved with them—as is the road to institutionalization.

Allow me to spin a sad tale of woe. Call it a Geek Tragedy. A visionary leader starts a business, and things take off. There's more work than people to do it, so more people are contracted or hired. The enterprise grows. Pretty soon enough money is floating around to support a much-needed back-office staff.

Things are indeed on the way up. Innovation is flowing. Really cool code and ideas are being executed. This place is a lot of fun!

And then something unexpected happens. A customer isn't satisfied. A deliverable is missed. Someone makes a mistake.

Such unexpected occurrences are usually the result of actions involving one or two individuals. This is all normal. Nothing bad has happened. Yet.

Reacting

And then management reacts. In and of itself, reacting isn't a bad thing. The path of least resistance, though, is to institute a policy. Please pay attention to the verb "institute." The policy, like all policies, is for everyone and forever.

Don't believe me? When was the last time you saw a policy removed? I thought so.

Policies are OK as long as they make sense for everyone and are applied to everyone in a uniform manner. In fact, my definition of a policy is something that makes sense for everyone. Things like "No weapons at work," or "No drinking on the job," are no-brainer policies for most occupations (with the exceptions of police officers and wine-tasters, respectively).

But what if a policy doesn't apply to everyone—or shouldn't?

Institution

Pull up a chair and sit a spell. Grandpa Andy is going to tell y'all a story:

There once was a company in Roanoke, Virginia that found their employees had been wasting time on the Internet. I would name the company, but I sincerely believe all publicity is good publicity and I want them to receive no benefit from this post. Management decided the best way to address this was to enact a policy and post the names of the web surfers and websites visited on the company intranet.

Do you know what happened next? The most qualified employees (arguably some of the best employees) found other jobs quickly, resigned, and left the company. The average employees found other companies to work for in time. The employees who felt trapped remained and suffered the indignity of this kind of treatment.

Those who remained spent a lot of their time looking busy. The last thing they wanted was more shame from management. Were they productive? No. They were spending all their knowledge-working cycles avoiding more punches to the brain.

That will teach those rascally Internet-browsing employees!

What Should Have Happened

Instead of lowering the bar, productivity, and collective IQ of the corporation, this company should have addressed the offenders directly. This thought would never enter the minds of a corporate management team capable of deciding to publish the browsing habits of their employees, but leadership should always seek to praise in public and correct in private. The price the company paid went beyond losing good employees. It also meant that innovation was stifled.

Does This Still Happen?

I don't know if the company in question operates in the same talent-hemorrhaging fashion these days. But these practices remain common in many organizations. Take a look at the marketing material from the many companies ready to sell you a solution to your employees' wasting time online.

What's the Big Deal?

I can hear you thinking, "What's wrong with companies controlling their employees' access to the Internet at work?"

Two words: trust and respect.

Everyone who thought that question needs to stop right now and read J. Ello's ComputerWorld article: The Unspoken Truth About Managing Geeks.

Done? Good.

Treating your team poorly does not produce better results. It produces worse results. It can mean the end of innovation. That means less bonus money for tyrannical management—to put it into terms they'll attend to.

Treat your team with trust and respect, and they will work harder and smarter for you. They'll produce more and better results, which will translate into even more bonus money for management. And, it's actually less work for management. So you can scale. Add more people. Produce yet more results. You get the picture.

It's tempting to institute policies, especially after you've been burned a couple times. It's certainly easier and simpler to institute policies than to change the way you manage. Is your goal to make your life easy and simple? Is that best for your company? Your customers? Your employees? You? Will the policies inhibit innovation?

Apply Andy's Policy Test™: Does your plan to make a companywide policy make sense for everyone? Or is it a matter best addressed with one person?

Chapter 18. Perfection versus Precision

This chapter is about aiming for the highest possible quality. A common complaint I encounter is that developers are perfectionists. Managers rarely come out and say that, but they convey that idea through their expectations and where they put their focus.

If you want to be able to plug-and-play developers like you would assembly-line workers, discouraging perfectionism is a time-proven strategy. But jobs that don't require a commitment to great results are all but gone from technical fields, and are disappearing from emerging economies.

If you are aiming at the middle, you will undoubtedly hit it. Aiming low is one interpretation of the consultant's mantra: Under-promise and over-deliver. If you aim low and deliver big, you look like a rock star. Aiming for the middle is safe.

But deliberately aiming for the middle eliminates the value of failing. Too many people underestimate the value of failure. You are not truly free to succeed until you are free to fail. Aiming high creates an opportunity to fail, which is an opportunity to learn.

And, every now and then, you hit your goal. Then what? Celebrate!

Truth be told, successful software development shops don't aim for the middle. There is a direct correlation between companies that produce innovative software and companies that are cool places to work. There is also a correlation between innovative software and teams that are made up of cool, quality-driven people. In fact, you can see great and innovative software from a great team at a very uncool company.

Semantics

Aiming for the stars, even if you sometimes fail to reach them, is a key to successful software development. Managers should cherish a developer who is a perfectionist.

Or maybe managers should think about the concept of "perfectionism" differently. Let me explain. When I find myself stretching to meet some goal—especially if that goal is delivering quality software—I am not practicing perfectionism.

Actually, I am practicing precision. My goal is stable, reliable software that exhibits expected and predictable behavior. I want errors that are reproducible and repeatable, and error responses and logging and externalization that follow a carefully crafted design pattern. I'm reaching beyond good and pursuing great. Why would you want developers who are *not* committed to going beyond the mediocre?

Chapter 19. Diversity

This chapter is about the power of diversity. I'm not talking about social diversity. I'm talking about diversity of thought.

There are at least two ways of looking at anything. For example, you can say groupthink has altered the course of history. That sounds positive. You can also say groupthink is responsible for atrocities against humanity that scar history. Not so positive. Both statements are accurate, but one is spin.

No one wakes up in the morning and says, "I think I'll suspend logic today and just blindly agree to whatever anyone suggests." At least I hope no one does that. Groupthink takes time and, like other forms of mold and decay, is an organic process.

However, if a bunch of people have similar training in how to solve problems in one domain, why are people shocked when groupthink occurs? For example, if you put a gaggle of MBAs in a room and present a case study they will reach a conclusion. (MBAs are uniquely qualified to deal with case studies because this is the chief mechanism used to train them.) Odds are that a group of MBAs will employ the tools of their trade and analyze the case study from top to bottom.

There will not be uniformity, but there will be consensus. In some instances, consensus is exactly what you want. In other instances, you may want more ideas on the board and more approaches than what business schools teach.

My point? In every project, there's at least one place where a less popular (or harder-to-sell) option is a—or even *the*—differentiator. If you have like-minded people attacking a problem, groupthink may prevent such an option from surfacing.

When you need creativity, you don't want groupthink. Diversity of perspectives and backgrounds is the key to keeping groupthink at bay.

Convenience is often the enemy of diversity. We look for people with certain skills to build our teams. We search for folks with similar backgrounds to our own—sometimes unconsciously. We want to work for a company of like-minded individuals. It's... well, it's comfortable.

But will we succeed if we're trying innovate when we're surrounded by people who have similar backgrounds? Will we grow if we join a group of people who all think like we already do?

I don't think so, and I can give an example of the power of diversity: The SQL Server community.

Diversity is one reason I love the SQL Server community. We have lots of diversity. Trust me. We have DBAs with experience in multiple platforms, as well as DBAs who've never executed a query outside of SQL Server. We have people who specialize in performance, high-availability people, business intelligence people, and storage people, to name just a few areas of expertise.

These are just our differences within the domain of SQL Server! That diversity is one of our strengths as a community. The sure sign that this diversity promotes innovation is the mutual trust and respect we all have, and that is evidenced in the way the community supports and promotes every member. The very fact that there is a community that is made up of people with so many interests and specialties attests to the strength of mutual trust and respect.

As the SQL Server community demonstrates, diversity of thought is a good thing. It's not always easy to manage, but diversity ultimately produces better results. By appreciating and fostering diversity, you can create a stronger team that will surprise and delight you and those they serve.

Chapter 20. Ringing

This chapter is about natural development cycles and the concept of "ringing." Ringing describes the curve of a waveform that occurs naturally, as the diagram below illustrates. It's called ringing because striking a bell is one of the natural places this curve appears.

Because ringing is a natural occurrence, it can help a manager understand certain phenomena in a team. For example, I see this curve in software projects in the form of code churn, bug rate and severity, and developer enthusiasm.

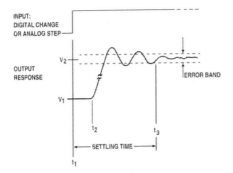

Code Churn

Code churn provides a generic way to measure the stability of software under development by capturing changes over time and across builds. It improves on the earlier lines-of-code-created (LOC) metric by measuring lines of code added, deleted, and edited.

You can see the flaw in measuring lines of code created in the following example: One of my more productive days as a developer resulted in removing 90 percent of the lines of code from a project I was working on (I was doing code-behind on a web page pre-.NET, before code-behind was automated). My LOC metric for that day was negative 1800, but the software was 10 times faster.

Code churn allows you to measure how the code changes as a way to understand the quality of the code as it evolves. At the beginning of a project, code churn will spike and bottom out as the development team figures out the best approach to solving complex challenges.

Once the difficult items are identified, it's common to start by developing those pieces of code. Solving the big mysteries first allows you to knock out a proof-of-concept and will drive the portion of the design that requires less problem-solving. It's not uncommon to develop several proofs-of-concept applications to test various approaches to the solution and even different tools. This makes for lots of code additions, deletions, and modifications early on. However, as development time nears the release date, code churn stabilizes to a predictable norm.

Bug Rate and Severity

When testing begins, you'll see lots of bugs reported and then fixed; then almost as many bugs reported and then fixed. The curve will ebb and flow, it's as natural as ocean waves washing on the shore. In general (and, of course, all generalizations are false), the number and severity of bugs will decrease over time. They will become fewer and farther between. Stability ensues as the ringing curve subsides.

Developer Enthusiasm

For developers, emotions ring at the beginning of the project until the unknowns are resolved. Developers need a challenge. Stop challenging them and they get bored. They respond to challenges differently, but most start new projects enthusiastically. Then comes the stark recognition of the conflict between the scope of work and the timeline. That recognition diminishes enthusiasm. As the project progresses, mysteries are revealed and problems solved. Gradually, the unknowns dissipate.

Ringing occurs in most projects, but it is normal. As a manager, it's tempting to react to the troughs or even the crests. Relaxing when things look good can be just as damaging as creating drama when things look bad. A steady hand is required to manage ringing. Realize it's normal, and you can help your team through the ups and downs of the curve.

Chapter 21. Passion

This chapter is about passion. Love and hate are not scalars; they're vectors. Hate is love moving in the opposite direction. The opposite of love is not hate; it's indifference. Indifference is also a vector, but its magnitude is zero. It's not going anywhere. If you're managing a team of developers, you certainly don't want hate, but you definitely don't want indifference or apathy, either. You want the team engaged. Developers tend to love what they do. They're passionate about their skill and the results they achieve. A team that loves what they're doing will be engaged because they have an intrinsic passion. As a manager, you have to recognize passion and understand how intrinsic passion and extrinsic motivation play together in the team. Otherwise, the things you may do to motivate the team could backfire and lead to apathy.

Engaged developers are passionate about their work. They have a sense of ownership, perhaps an overdeveloped sense of ownership. They work like crazy to deliver. And they take pride in their work. This all sounds great because, well, it is all great.

But the way developers communicate their passion varies. Most are protective of their code, to a fault. This means they will turn on you in a heartbeat if you criticize their baby. Some react poorly. Others behave badly. It happens.

You can write such developers off as poor communicators if you want, but this would be a mistake. Why? We admire the protectiveness of developers when they're busy delivering; it's an expression of their passion. But you have to take their sensitivity to criticism with that passion. It's a package deal.

In Community

Because they share a passion for their work and the pride they take in it, developers relate to other developers. They're members of a developer community, or tribe. The developer community provides motivation and strength to its members as they interact with each other.

In two years of managing teams, I've learned that developer passion and that sense of belonging to a community, or a tribe, give me a strong tool for motivating and supporting the developers I manage. To help me with this, I've studied a lot of concepts about how the idea of tribes applies in business.

One of the things I've learned is that some things that work in business can translate poorly if you try to apply them to community. I've come across some fascinating thinking about why that is. I've been reading <u>Tribes</u>, <u>Drive</u>, <u>The Cathedral and the Bazaar</u>, <u>Thinking Strategically: The Competitive Edge in Business, Politics, and Everyday Life</u>, and <u>Cognitive Surplus</u>.

More than one of these books offers interpretations of the work of <u>Edward L. Deci</u>, which supports incentive theories and the relationship between intrinsic and extrinsic motivation. That passion that developers exhibit is an example of intrinsic motivation. Extrinsic motivations include salary and recognition. Both are necessary, and you have to be aware of how intrinsic and extrinsic motivation affect each other and the developer you're working with.

One interesting tidbit about the transition from intrinsic to extrinsic motivations is that it's a one-way trip: When external motivations replace internal, the activity, interest, and motivation decrease: To explain what that means, let me give you a quote from <u>Self-determination theory: Basic needs and intrinsic motivation</u>:

> "Deci (1971) found that offering people extrinsic rewards for behaviour that is intrinsically motivated undermined the intrinsic motivation as they grow less interested in it. Initially intrinsically motivated behaviour becomes controlled by external rewards, which undermines their autonomy."

Salary?

You could read into Deci's statement the idea that an extrinsic motivator like salary isn't important to a developer who's passionate about the work. Using that reasoning, a lot of people will tell you this means salary doesn't matter. Ironically, those same people are likely attempting to increase their bonus or salary by controlling the salaries of others who work for them. <Sarcasm> It's a stretch, but I consider this situation a conflict of interest. <Sarcasm>

My opinion is the opposite of those who say salary doesn't motivate people who are internally motivated. However, in a conversation about passion and motivation, salary is a distraction. Like oxygen, salary isn't the issue—so long as you're getting enough.

The real issue is passion. If you want great results from developers, recognize and foster their passion. Too often, developers are told to temper their passion in order to gain extrinsic rewards. What have we done? What were we thinking? Oh that's right—it's more important to play nice than... what? Compete successfully? Deliver? Remember: Passion and developers is a package deal.

I see companies and communities struggling with passion. I see some figuring it out, and some of these are learning the hard way. Kill the passion, and maybe everyone feels good about their extrinsic rewards—until the company or community folds, that is.

Tolerate... No, *embrace* the passion, and everyone feels good about results and delivering and succeeding. The intrinsic rewards will keep your team of developers motivated and happy.

Chapter 22. Business Losses and the Power of "I Don't Know"

Sometimes it's important to say, "I don't know." When I hear someone admit that he or she doesn't know something, I gain trust in that person's integrity. I get suspicious if a person has all the answers or a company always wins. It's just not natural.

Take job interviews, for example. As a hiring manager, I check for lots of things when I interview people. Since I hire ETL developers, one thing I check for is SSIS experience. I have a list of SSIS technical questions I ask. They range in difficulty and help me check for the kind of experience an SSIS developer possesses. One of the answers I'm listening for is, "I don't know."

Why is hearing, "I don't know" so important? No one knows everything. (Well, maybe someone somewhere does. That person is definitely not me.) When I'm interviewing people, I expect to encounter a topic they are unfamiliar with. Since I know the answers to the questions I ask, I can tell when they know the correct answer and when they do not. If they take a stab at it and miss, I know I'm dealing with someone who is either misinformed or guessing. I ask good follow-up questions. If I do not get an "I don't know," I realize I'm dealing with someone who does not want to admit they don't know something.

What does it mean if you won't admit you don't know something? It could mean a lot of things. For me, it means you're not a good fit for my team. Why? Because we occasionally face situations where accurate information is vital. I liken these situations to man-overboard drills. If Harry fell off the ship and can't swim, I need Harry or someone who knows Harry to say, "Harry can't swim."

Once Harry is in the water, this is not the situation to teach Harry the finer points of dog-paddling or treading water. Before the splash is the time to find out if Harry will be able to make it until we turn the boat around, or if we need to send someone over the side immediately to help him remain afloat until we do.

If you don't know and won't tell me, you will slow us down. On my team, saying, "I don't know," doesn't count against you. It's an incomplete, but honest, answer. When members of my team say it, I know they mean it.

In my presentation "Database Design for Developers," I point out that the only wrong answer is, "I don't know." It's wrong because we're IT professionals and are paid to know. However, if your choices are lying or saying, "I don't know," don't lie. Say "I don't know."

Saying, "I don't know," tells me that the person I'm talking to is able to admit an uncomfortable truth. Admitting an uncomfortable truth is also an important criterion for winning my trust when I'm dealing with a business organization.

For example, when I was a consultant I used to interview with companies. My interview was usually with the individual who would serve as my point-of-contact throughout the engagement. Near the end of the interview, the person I was interviewing would usually ask me if I had any questions. I always asked, "Have you ever lost money?" If they said "Yes," I asked for non-specific details. I didn't want to know how much or to whom; rather I wanted to know that they'd done business and circumstances hadn't always gone in their favor.

This is important. It's similar to the I-don't-know points above. If I'm about to engage in business with people who always win and never lose in business, I feel that they're either not admitting an uncomfortable truth, or they conduct business in a ruthless fashion.

I don't do ruthless—at least not in business. The world is not so cutthroat as to require businesses to believe it's OK if they thrive at the expense of others. I've been pretty hungry before, but never so much so that I intentionally engaged with people who believed they win when I lose.

I'm not advising you to run out and burn good money so you'll be able to answer me by admitting you've had a contract that went south. I am advising you to admit your failures and realize that failure is part of doing business.

Beware of doing business with individuals who know everything and companies that always win. Trust me on this. There's a reason. Whether it's hard to admit a failure, or whether an entity will do whatever it takes to keep from failing, knowing everything and always winning is not feasible or sustainable.

Chapter 23. Personality Clashes, Style Collisions, and Differences of Opinion

This chapter is about how business people and developers interact. In the context of developing code, sometimes business people don't know what they want. There's nothing wrong with this. In fact, it's more often true than not true. Agile development methodologies are based on a recognition that *not* knowing the requirements when a project starts is normal. The problem arises when business people who don't know what they want try to interact with developers who want to give them what they want.

Such situations tend to bring out the ego in both the business person and the developer. Every recipe for disaster includes the line, "Add 1 ego." Egos are great sources for problems that are actually self-inflicted wounds. Egos are usually a symptom. They indicate fear. Fear of what? It could be anything, but most likely it's fear of what others think.

Egos just don't work well in our field. Egos are a source of pride and pride is the root of all kinds of ugliness. I try to live by Romans 12:3:

> "For through the grace given to me I say to everyone among you not to think more highly of himself than he ought to think; but to think so as to have sound judgment, as God has allotted to each a measure of faith."

I'll let you in on a little secret: Whether you care what others think or not, you can act as if you do not. The way you behave tells others how to treat you. Insecure people will take advantage of you if you let them. It's what they do. Don't let them.

One Solution

One way to address personality clashes, collisions of style, and differences of opinions is to communicate. <u>Drive: The Surprising Truth About What Motivates Us</u> is one of the best books I've read. It helped me with my business communication because it focuses on the things that motivate us, and those things are actually *behind* what we're trying to communicate.

Because most business people do not know everything they want when the project starts, one valuable skill database professionals and software professionals can cultivate is communication. Communication is really a two-phase commit that consists of the message-sender initiating communications and the message-receiver interpreting the message. What is actually communicated is an interpretation of the received message, which may be wildly different from the intended message from the sender.

A good story (or storyboard) can introduce the developer's vision to the business person. User experience tools can provide this capability, as does Visio, or Paint, or even a cocktail napkin. The point isn't the tool used to communicate; the point is to *close the feedback loop by showing the business person how you interpreted the message*!

When you remove ego and communicate with each person's motivation in mind, you can overcome barriers and ensure that the business people get want they want. Developers can feel successful by knowing what is expected and delivering it.

Chapter 24. Human Resources

This chapter is about how I feel about the term "Human Resources." I think it's a bad name for a function that deals with people. It conveys an inaccurate understanding of the nature of employees and their importance to business.

When I say I dislike the Human Resources, I don't mean the people who make their living doing Human Resources jobs. I don't dislike the people. I dislike the idea for three main reasons. First, the term "Human Resources" is dehumanizing. Second, the term Human Resources implies that people are a commodity, a "resource" to be used. Third, Human Resources squashes motivation. Let me dive into each of these reasons.

Human Resources Is Dehumanizing

Using "dehumanizing" to describe a two-word phrase that includes the word "human" may be ironic, but the problem is not with the word "human." The problem I have is with combining "human" with the second word in the phrase, "resources." "Resources" doesn't belong with the word "human." People are in their own category; separate and apart from resources.

The definition of "resource" highlights why I believe it's inappropriate to call people "resources." A resource is:

1. Something that can be used for support or help: The local library is a valuable resource
2. An available supply that can be drawn on when needed. Often used in the plural
3. The ability to deal with a difficult or troublesome situation effectively; initiative: a person of resource
4. Means that can be used to cope with a difficult situation.

Resources are things that humans use. But resources are not people. People should not be treated as "an available supply that can be drawn on when needed." That definition implies that people are just around to be used as needed and are expendable and replaceable, like other commodities.

When I think of resources, I think about baking. I am married to a person who bakes. Christy is an awesome cook and baker. She uses resources when she bakes. She has resources that include cookie-cutters, rolling pins, and fancy work-mats that keep her ingredients dry, malleable, and otherwise ingredient-ly.

She can easily replace a cookie cutter if it breaks (or, hypothetically, if one of the children uses it to make mud pies in the garden and leaves it there over the winter to be discovered when it's hit by a tiller blade in the spring. Hypothetically).

But people are not like cookie cutters. Yes, you can replace people. But it *hurts* when someone leaves a team. Finding another person is hard enough. Finding someone who *fits*? Extremely difficult. In contrast, replacing a resource is easy.

The term "resources" is confusing at best and just plain wrong at worst. It reinforces the idea that you can unplug one person and plug in a new one. This is the thinking that pervades the myth of "savings" with offshoring. It fuels the race to the bottom.

HR Flattens

Let me return to the baking metaphor. Human Resources rolls us all out like dough, aiming for a consistent texture of qualifications. Standard Human Resources procedure measures everyone against a set of criteria that is subject to the requirements of an hour, day, week, month, quarter, or project.

.Standard measures can't account for changing needs. The qualifications required for any job don't just shift; they *always* shift.

What HR is trying to do is admirable: Apply scientific principles to human behavior to ascertain who the best fit for a team is. But true scientific principles take deviations from the standard into account, and Human Resources does not.

Human Resources Is a Motivation Sink

By focusing on extrinsic rewards such as salary, standard Human Resources practices can actually suppress intrinsic motivation. If a person loves his or her job and the only way to progress in his or her career is to strive for promotions and more money, that person is forced to focus on career rather than the joy of doing the job itself.

Human Resources is an organization's source of tools and advice for managers to help their teams succeed. Ordinarily, the tool Human Resources provides is extrinsic rewards. Some argue reward is the goal of performance management.

I think you need to consider two key things when you think about rewards: First, rewards are not just about money and career. Second rewards don't happen in a vacuum. Let's look more closely at these two areas, keeping in mind that they are difficult to balance and easy to mess up (with tragic results).

Not Just Money

Some business people read my claim that rewards are not just about money and get all excited. Such people think, "Andy is going to tell me the magic formula to pay my folks as little as possible and still get the maximum of work out of them!"

Let me stop right here. If you had any inkling of that thought, then you likely believe that reward *is* all about money. You need to pay your people more. How do I know this? Because if you had that reaction, you're a stingy boss who has little business managing a team. That's how I know.

Pay is like oxygen: It's not a problem unless you don't have enough. If you want the absolute in mediocre performance out of your team, pay them the industry average. The converse applies: If you're paying the industry average (or a smidgen less because times are tough and you believe they need this job more than you need them), you're getting what you're paying for.

"But Andy, I've heard your presentation on managing teams, and you reference all these studies that say money won't motivate people to do more."

You're correct. And people are standing outside every business school on the planet shouting that theme *before* you get into class on day one. Here's what business schools understand: The upside for motivating people with money is asymptotic. As the monetary reward increases linearly, performance increases in ever-smaller increments. There is a limit on the upside for financial reward.

Here's what business schools are missing: The downside effects of not-enough-pay are not simply the inverse. It's a pivot of the asymptote. Demotivating someone by not paying them enough is a surefire way to send them on a quest to see just how little effort they can put in each day before they hit a point management recognizes as underperformance.

Underpaid employees become a management nightmare, consuming the manager's time. The manager must either constantly monitor or fire that employee and then embark on the search for another ~~hapless victim to underpay~~ employee.

If you do the math, it's often *less expensive* to simply pay the employee fairly. The value is high in an employee who knows the business processes, holds institutional knowledge, has built relationships within the enterprise, and understands your customers.

Once pay is satisfactory, you'll find other ways to reward employees. But not before *pay is satisfactory.* Training is one example. I saw a tweet that said something along the lines of: "What if we train our people and they leave?"

An answer to ponder is: What if you *don't* train them and they *stay*?

Technical specialists crave and appreciate training. Technology is changing at an ever-increasing pace, so training is required if technical people are to keep their skills current and relevant.

Training your employees communicates that you care about each employee's career. Demonstrating that concern will foster loyalty. Will all employees that you train stay with your company? Heck no. And that's unrealistic. But more of them will stay, and they'll know more and do more and better work.

You wouldn't bother training "resources." Resources are simply things you can use and are replaceable. Not training employees communicates to them loud and clear that the organization would train them if the company were going to keep the employee around for the long term or cared about the employee as an individual.

Let me give you an example. A good friend just changed jobs because the new company guaranteed he would be able to attend the PASS Summit each year for training in his specialty, SQL Server. He had been at his old job several years, had been available to support deployments and production at all hours of the night. He is as honest as they come. Losing him will cost his old company a lot more than the $2,000 or so it would've cost to send him to the PASS Summit each year. To the old company, he was just a resource.

Other ways to reward employees include flex-time and working remotely. Both require management to trust and respect their teams. For some managers, this requirement of trust is asking too much. Their attitude implies that *resources* don't have lives outside of work.

But people *do* have lives and families and hobbies. And just other stuff that they would enjoy doing. Shocking, but also true. Managers who acknowledge and respect their team members' personal lives gain good will and loyalty from their teams.

Reward Doesn't Happen in a Vacuum

Wouldn't it be nice if we could compartmentalize life? If every action didn't have an equal and opposite reaction? I have to admit, that would come in handy sometimes. But the universe isn't wired that way. If a manager thinks it's OK to throw an employee an occasional reward while ignoring the employee's personal needs and intrinsic motivation, that manager will fail to get lasting performance. The employee who is treated as a resource will behave like a resource.

Management is about knowing who you, the manager, are as a person. But it's also about recognizing who your team members are as people and treating them with the respect they deserve as humans.

Chapter 25. Sounds Good

This chapter is about consequences. Ever buy a house? Or a car? Or an idea for a new feature in an application or website? What these have in common is that they require decisions that have consequences. Depending on how well you consider the decision, you can end up with both intended and unintended consequences.

No decision occurs in a vacuum. Every decision you make in favor of Option A is a decision against Options B, C, and D. This elimination of some options in favor of others is part of the physics of a decision, business or otherwise. As the table below shows, the consequences of choosing some options and eliminating others can be positive and negative.

	Positive	Negative
Intended	Intended Positive Consequences	Intended Negative Consequences
Unintended	Unintended Positive Consequences	Unintended Negative Consequences

When making decisions, you have to try to foresee all the consequences and make sure you're minimizing the unintended ones. The substance of decisions is unintended consequences. Unintended consequences leave a mark. They are merely lacquered with intended consequences. This is why unintended consequences make such wonderful paving stones for the path to hell.

Unfortunately, it's often easy to make a decision without fully considering the consequences. The path of least resistance is compelling because it's easy. But easy ain't always right. I'm suspicious if the first idea for a solution or the first impulse for a decision is easy because I rarely get things right the first time. The easy first thought is often incomplete, or doesn't address all the need or solve all of the problem, is not elegant, or creates additional problems. The first thought lacks vetting. I haven't had to consider the unintended consequences. It is therefore suspect.

Unintended Management Consequences

Challenging the easy answer is important in managing teams. When I make a bad decision on a car or house or personal website feature, I'm mostly hurting myself. But there's an added dimension to making a boneheaded management decision: I can hurt someone else.

As a manager, I've done this. I've made bad decisions that cost effort, time, and money for me and those I serve (I say, "those I serve," because I believe that management is service). Nothing stings like hurting the team. Inevitably, those costly decisions have been the result of my taking the path of least resistance—like when I've gotten upset at what I saw or heard, or read, without taking time to investigate or digest something and felt froggy and jumped all over someone who was actually trying to mitigate bad consequences for me and the team.

Bad for me? That I can handle. Bad for members of my team? That just stinks.

Patterns vs. Exceptions

We all have patterns in our lives that keep us repeating mistakes. But if we're self-aware, we can catch ourselves in those patterns and create exceptions to the rules of our patterns. If my mistakes as a manager have any redeeming quality it's this: They are the exception and not the rule.

It's easy to get into an unhealthy pattern of management. We have the perfect excuses: It's a tough job; team members alternate between expecting too much and too little from us; no one appreciates the 99 things we successfully fend off, but everyone notices the one thing that gets through. The front-line manager position is not for the thin-skinned or faint-of-heart.

If management were easy, anyone could do it.

Serving

So how do you catch yourself in your bad patterns and avoid unintended consequences? Think about what your role as a manager really amounts to and act accordingly.

As I said, I believe that management is a service. In my experience, the best managers sacrifice for their teams. They work hard and smart, refusing to select one over the other. They often put in as many hours as their team members or more. They rarely perform the work that keeps the machine running, but they protect the environment of those doing the actual work. And they remove obstacles for the team.

Do you think serving a team is a small task? Here's some perspective for you: It takes something like six support personnel to support each infantry person in an army. Want to fight to win? You can't do that by running onto the battlefield, firing a couple shots, and then retreating. You have to project power to win a sustainable (and honorable) victory. Projecting power means logistics, supply chain, support, and so on. In other words, management.

The example of the infantry is on point. It's a job that the protected cannot pay the infantry enough for. God bless them. But they can only carry so much ammunition on their person; they need someone to supply more. Their boots wear out. They need to eat. You get the picture. They need someone (or a group of people) to facilitate the environment; someone to enable them to do their job.

Similarly, a team needs a manager to stay ahead of the team's needs and make sure they are equipped to do the job. A manager has to know what the team needs, make decisions that are in the team's best interest, and ensure that the consequences of management decisions are thought through to avoid unforeseen negative consequences.

Chapter 26. Coopetition

This chapter is about competition in a connected business climate. I call it "coopetition." It's when you work hard to win, but not at the expense of relationships and good will. Coopetition means that at the same time you compete with others, you also find ways to collaborate for the greater good. Coopetition can be within your organization, as well as outside it, involving other organizations or individuals who ordinarily compete with you for jobs, or resources, or attention. By opening yourself to opportunities to collaborate, you create business relationships that are sustainable and can take you to unexpected wins.

How You Play the Game

Someone comes out on top in every contest. There are a few ways to think about the contest and the outcome. Several interesting ways are described by examples of Nash equilibria (named for John Forbes Nash, Jr., a winner of the Nobel Prize for Economics and subject of the book and movie *A Beautiful Mind*). But there are other ways to think about these things: short term, long term, and sustainable.

Short Term

One way is short-term. The idea is the whatever-it-takes-to-win contest. Rules exist to be stretched and circumvented. Vote early and often. Coerce the panel. Stack the deck. Winning is the most important thing.

To those who embrace this type of competition, I'd like to say, "Welcome to the 1980s." I'd also like to ask, "How are you sleeping?"

We all participate in communities. If you choose not to participate in a community, you are *still* participating in a community, just not very effectively. You're also living in denial, but that's not your most important issue currently. On some level, you're dismissing and writing off your community. On another level, you're disconnected. Like it (accept it) or not, social currency, exists, and you're running a tad low.

Long Term

People who think long-term consider a bigger picture. They're in it to win the war, not every battle. They choose their battles carefully. By carefully, I mean strategically. As a result of thinking strategically, they accept the occasional loss to achieve a greater goal. There are hills worth dying on, but those hills are few and far between. Better to live to fight another day.

Sustainable

I do not like fighting metaphors when it comes to describing sustainable competition. It's because war is not sustainable. Someone eventually loses. Truth be told, everyone involved in a war loses something. While I believe some causes are worth fighting over, I'm not convinced war is the best way to wage the fight. I see it instead as the last option. Sustainability is about making something work for the long haul.

To accomplish sustainability, a positive spiral is required. What's a positive spiral? There are lots of definitions and theories. My favorite definition begins with stasis and the status quo. Things are humming along at a decent clip, nothing too bad is happening, but nothing too good is happening either. And then along comes a disturbance, a disruption. Perhaps some apple carts are upturned, maybe some feathers are ruffled. But the disruption stirs the soul of the organization (or business, or individual), and before you know it, something good has come from it—a positive spiral emerges, a sustainable bias towards the good.

Demonstrating you'll stop at nothing to crush competition is not sustainable. The hidden message you send to your team and the market is: "I'm a selfish and insecure individual/business/organization. I cannot tolerate others in this space."

You may think everyone will want to work with such a big winner. You would be wrong. If your ego is *that* destructive, it will soon run out of external targets. At that point, little is left but internal targets, and you will begin taking your organization apart one snide comment at a time.

There's Enough Work to Go Around

Whether the question of collaboration versus competition comes up in connection with another individual in your work environment, with another company talking to the same pool of potential customers, you can find ways to work together for the greater good. There's plenty of work out there for everyone. In fact, you can even earn a percentage off of work you send to competitors. How's that for a win-win?

And let's not forget about good will. Or bad will. Good will is one more reason people will choose you over your competition. Bad will is a reason for people to seek out and choose your competition. A common mistake of de facto monopolies is to think they must be doing things right because there's no competition. Sometimes, the bad will is accumulating and will convert into momentum for competition, once it arrives.

If you look for ways to serve your team, your organization, and your community, you'll find that you'll have a sustainable model of coopetition that's good for everyone.

Chapter 27. Disruption

Disruption requires disruptors. Most organizations have a sign out front that reads:

> NO DISRUPTORS ALLOWED!
> THIS MEANS YOU!

Does this sign physically exist? Of course not. It's there, though. If you want to work in an environment that doesn't stifle creativity and new ways to approach problems, think about the following: Check out the hiring policies. Take a gander at the pay. Look at the process for nominating members of leadership. Is there a plays-well-with-others requirement? Does the process scream, "Same old, same old"? Or does it convey "Innovation Welcome Here!"

Change Hurts

If you're an agent of change, or a disruptor, stay away from organizations that fear change. Change hurts, but it is necessary for long-term success. You need to be sure that the people you work with are willing to examine their motives and actions to make sure they're moving in a positive direction. Remember that change never occurs in a vacuum. This includes changes in thinking and mindset. If people do the right things for the wrong reasons, the results will not be sustainable.

One important change that's hard for companies and individuals involves how you think about your competition. Growth comes out of changing from thinking of your competition as your worst enemy to considering them the entity or individual that will willingly show you your weaknesses and flaws. Knowing yourself is the first step to success.

The process of change isn't easy or fun, but the alternative is disastrous, whether you win or lose. But how you go about change and doing business with others can make all the difference between a change gone wrong and a change for the better.

Allow me to demonstrate: Suppose you implement a change that causes you to lose. That's easy right? Your business model collapses, business stops, customers go elsewhere. All bad and all fairly simple to understand. But where would you have been if you hadn't tried to change?

Suppose you win. In this case, a lot depends on *how* you win. If you are completely above reproach in the methods you choose for competition, your ethics spotless, your actions unimpeachable, then you're probably fine.

But if you weren't ethical, the win may be fleeting. Nothing happens in secret. Stuff that happens in Vegas does not always stay in Vegas. People remember. If you cut a partner, every other partner will harbor suspicions that you will do the same to them. If you've done it before, you have demonstrated you're capable of doing it again. If you engaged in business for the sole purpose of destroying competition, you've demonstrated to your team and the market that winning is everything to you, regardless of who gets hurt. This is especially true when the "winning" entity is large and established and the "loser" is not.

A gut check is required when you enter into change. What are your motivations? Are you doing what's right, or what's expedient? Real, live change is required to reverse negative patterns of behavior. Not lip-service. Or doing enough to make people think you're changing. What people think will not take you down; only *you* are capable of that. A change that means doing what's right will be disruptive.

You will not like it. You will like not doing it even less. But a change towards what's right will have a positive effect on your business, your relationships, and your career.

Chapter 28. Recognition versus Satisfaction

The theme of this book is about knowing who you are and doing what you are in order to be successful. As I've said throughout this book, my motto is, "Be who you are; do what you are." Sometimes I need to be reminded about that principle and make sure it's guiding and motivating me. This is a story about how someone I deeply respect straightened out my perspective.

I was nominated to receive Microsoft's prestigious Most Valued Professional (MVP) award in 2006. I was nominated but didn't receive the award. I remember talking to The Best Damn Developer Evangelist On The Planet (TBDDEOTP) G. Andrew Duthie about it.

Andrew asked me point-blank, "Andy, why are you serving the community?"

It was an in-your-face kind of question, and it had the desired effect. After collecting my thoughts, I replied, "I'm serving the community because others served me. I learned from them, and now I want to give back."

Andrew's response: "Keep doing that. Recognition may or may not come. But you already have satisfaction. Look at the lives and careers you're impacting!"

(I told you he's TBDDEOTP!)

I was nominated again in 2007 and was awarded MVP status. But Andrew was and is right: What I do can't be about recognition. Recognition is not fulfilling in and of itself. The victory is fleeting and the triumph hollow. You have to have a good reason for wanting to reach the goal. You have to do things because they are what you are.

What Andrew did for me was to make me think about my *purpose*. Purpose denotes more than a mere goal, such as achieving MVP status. Goals serve purposes, such as giving back to the community that has served me. Purposes impact lives, careers, and communities. Purposes help people! Helping people continues to be satisfying long after the thrill of recognition.

Andrew also reminded me that you have to be willing to try, fail, and try again. And again. And again. You need to be able to harvest the lessons from the last outing—the good and bad, the flattering, and humbling—and repurpose those lessons into the next venture. When you accomplish something because it grows from who you are, you'll find that it's more satisfying than any award.

Get. Started. Now.

Are your ideas going to be awesome and make you a million bucks by the end of the year by doing what you are? Probably not. Overnight success takes about a decade. But you will never reach your potential by striving for external rewards. Take time to think about your purpose rather than what you get. Do. Something. Now.

The odds are you will fail, as I did with the MVP award. That's a good thing. Have you ever been around someone who never fails? Something is fishy about people who never fail. Failing prepares you for success. It's a prerequisite. But failure can be the start of achieving true satisfaction.

Chapter 29. Do You Have a Job, or Does Your Job Have You?

Do you have a job, or does your job have you? It's an important question and distinction. If you have a job, your contributions should be valued and some of them (at least) should be implemented. If the job has you, the company probably just needs warm bodies and yours is as good as almost anyone else's. By asking if you have a job or your job has you, I'm stopping by your life and asking, "Who's in charge here?" Is it you? Is it someone else? Do you even know?

So how can you tell if your job has you? Let's start with some more questions:

1. When was your last vacation? According to a recent study, "Americans collectively failed to take more than five hundred million (577,212,000) available days of vacation." Are you one of those people? Ok, wait.

   First let's define vacation. You don't have to travel to have a vacation. But you do have to be 100% completely in control of your time with 0% chance of work calling. You can take off your watch (if you wear a watch; I don't). You can double-Ziploc-baggie your cell phone, put it in an old cigar box, and bury it out back until you return to work. You can leave an Out Of Office message that reads "I'm on vacation. If this is an emergency, please call 911. Then email me the details. I want to read about it when I return." That's a vacation.

2. When was your last raise or benefit increase or quality-of-life-improving recognition? If your answer includes the word "year," that's a clue. If we're talking about money, the raise has to be more than inflation. But the important factor doesn't have to be

money; it can be training or the opportunity to attend community events with your expenses paid.
   3. How many of your ideas have been implemented in the past 12 months? You don't have to be in product development to introduce efficiency and solve problems. You can suggest implementing a product or process that improves the efficiency of SQL Server backups. The heart of this question is really: "Is your employer taking you seriously?"

Why These Questions Matter to You

Vacations: People need time away from work. It doesn't have to be two weeks in a foreign country, but it needs to be a break from work. It's good for you, and it's ultimately good for your employer. If you have a job, vacations are part of the plan, and your employer not only encourages you to use your vacation time but also makes sure you have the support to be detached from your phone and email. If the job has you that may not be the case.

Rewards: Compensation needs to keep up with inflation at a minimum. It could be that you're working for a company that is going through a tough time and is therefore unable to give raises. The company may very well make up for lower pay later when finances improve. Another possibility is this may not be the company, position, or field for you. Feeling like you're stuck in a company that doesn't care about fair compensation is a sign that the job has you. It's not the end of the world if you need to find a more satisfying organization or career. It's the beginning of a new road. If you have a job, you can have another. If the job has you, it's difficult to make this kind of change.

Employers should not underestimate the value of training for the recipient and for the company. But sometimes managers are threatened by helping employees become more qualified. I'm reminded again of that Twitter exchange to the effect of, "What if we train our people, and they quit?" What a sad mistake. My response is, "What if you don't train them, and they stay?" If your company is threatened by your desire to improve your skills and you just accept that attitude, that's a sign that your job has you.

Suggestions: If your suggestions are not being implemented, the possibilities include:

1. You're not contributing ideas. Start contributing
2. Your ideas aren't very good. Learn more about your field so your ideas are informed and relevant
3. Your good ideas are being ignored. Talk to your manager and your manager's manager and find out why
4. Everyone's ideas are being ignored. The company does not respect its employees or is threatened by change. Find a new employer

As a manager, I've learned that my career is in my hands, and I refuse to accept a job that wants to have me. Because these things are important to me, I make sure my team gets their vacation time; receives appropriate pay, training, and recognition; and sees their ideas implemented (or understands why ideas are not implemented). I've also learned to encourage the people on my team to change things that make them feel their job has them.

Chapter 30. Conviction

As a person who believes in doing what I am, I find that standing up for my convictions is one of the most important aspects of being a good manager. Conviction has several meanings, according to the dictionary:

1. Firmly held belief: a belief or opinion that is held firmly
2. Firmness of belief: firmness of belief or opinion
3. Guilty verdict: an act of finding somebody guilty of a crime, or an instance of being found guilty

This chapter is about a firmly held belief and firmness of belief or opinion. To act on what you are—as a person and as a manager—you have to understand conviction and the importance of acting on it with balance and openness.

Firmly Held Belief or Opinion

People occasionally confuse slogans or clichés for firmly held beliefs, or conviction. They hear something repeated so often that they start to perceive it as reality. (Note the two platitudes I included in that last sentence: i.e., "Perception is reality," and "Repeat a lie often enough and people will believe it.") I call this phenomenon of mistaking slogans for conviction succumbing to groupthink.

Groupthink is dangerous because it means no questioning is going on in the minds of those who take platitudes at face value. That's the opposite of a conviction, which requires deep reflection and even challenging the ideas underlying a belief or opinion. When people don't examine the basis of their opinions and simply accept groupthink, they can be easily manipulated and goaded into attacking others who do not subscribe to the same groupthink.

It's one thing for good people to disagree. It's a completely different thing when groupthink kicks in and the conversation degenerates into a repetition of slogans. Repeating slogans is also, well, childish. And dangerous. When you really examine what's behind resorting to slinging slogans, it's as close to an admission of failure as you're going to get. If no real conviction is behind groupthink, all that's left is the empty platitudes. That means the groupthinkers have lost the debate because they don't have a defensible position and have no choice but to devolve the conversation by using schoolyard tactics and bullying.

In contrast, people who have examined their beliefs and opinions are open to challenges and can use reasoning and intellect to have a true conversation with someone who disagrees. The discussion may not lead to agreement, but it will maintain good will on both sides of the issue because empty slogans and groupthink are not involved.

Firmness of Belief

Conviction can also represent a strong belief. As I write this, I'm watching the movie *Iron Man*. Interestingly enough, it's at the scene where Tony Stark says, "I know in my heart it's the right thing to do." That's the kind of conviction I'm talking about here. This type of conviction drives your actions and holds you accountable to an internal code of conduct.

If anything, conviction (in the strong-belief sense) is about as anti-groupthink as it can be. It's isolating because it's about your own moral center. This type of conviction requires openness to risk and uncertainty because you're standing for something that makes you behave in a particular way—and not everyone will agree with or accept that behavior. In the end, you'll either be despised as mad or heralded as visionary. There's little middle ground, but that's because straying from your conviction means that it's not really a conviction at all.

Balance

In 1854, Henry David Thoreau wrote Walden. Thoreau famously wrote, "The mass of men lead lives of quiet desperation. What is called resignation is confirmed desperation." Some folks hold deep convictions but have resigned themselves to never acting on those convictions due to circumstances in their lives. These people are examples of what Thoreau meant by living lives of quiet desperation. Thoreau advocates recognizing that resignation and overcoming it. Acting on convictions is what gives meaning to life and releases people from those lives of quiet despair.

Of course it's possible to choose to act out of proportion, to overact on a conviction. But then the danger is straying into groupthink and unthinking sloganeering.

The key is balance, but balance is sometimes hard. Balance is difficult, especially when you hold a conviction and are in the minority. The way to achieve balance is to allow yourself to question and sometimes doubt. You think you're right (obviously), but are you sure? This is when you can draw on conversations and input from others. You might end up confirming your convictions, and you might end up finding new dimensions to explore.

Courage

Holding and acting on any conviction, but especially a minority conviction, requires courage. Courage does not mean the absence of fear. It means acting on conviction in the face of fear but being open to learning. If it was easy anyone could do it.

It's not easy.

Thoreau's admonition to overcome resignation and move beyond quiet despair is a lens for focusing in on your convictions and acting on them. The Bible also sums up conviction: "Be strong and courageous. Do not be afraid; do not be discouraged," Joshua 1:9.

This is as true in management as in life. Your values inform your leadership and are reflected back to you by your team. Examine your convictions and stay away from unchallenged slogans and groupthink.

Chapter 31. I Type in Real Time During Demos

My friend, SQL Server guru and highly respected speaker, Buck Woody advises, "Never type in demos." But I make a point of typing in real time rather than prerecording the keystrokes. I'm not typing in demos to irritate my friend. I love Buck. He rocks. I type in demos to fail.

When you give a presentation, attendees are in the room for a reason. Usually, part of that reason is to learn something new or to find out more about a technology or business practice. But in-person events provide so many more opportunities: meeting new people, networking, sharing a meal, looking people in the eye, etc. A major opportunity is that an in-person presentation gives you and me the opportunity to take a technological journey together: "I'm Andy, and I'll be your tour guide today as we navigate the sometimes perilous river of the SSIS Expression Language." Typing in a demo is one way to convey that we're in this together, and we'll figure out the obstacles as they arise. Failure, even in a live demo, is a way to learn.

If I try and fail, what really happened? Some will undoubtedly view the presentation as worthless, delivered by someone who can't even get their own demo code to behave correctly. There's a possibility I will alienate attendees who feel this way.

I think the positives outweigh the potential for alienating some people so much that it's worth taking the shot time and time again. Let's look at what those positives are.

The Positives

I've been working with SSIS for a long time. Before that I worked with T-SQL, Data Mirror, Business Objects, Visual Basic, and a few other programming languages (starting in 1975 with M6800 machine code when I was 11). I have experience writing code.

"Oh, so you're showing off," you might be thinking. No. I'm showing on. I mention my credentials to show that no matter how experienced I am, I can always find ways to fail—and learn from those failures. When I'm demonstrating software or some technique or design pattern, I'm demonstrating that I can and sometimes do fail. I'm also demonstrating risk is not to be feared, shunned, or accepted as an excuse. Risk is to be tested.

My hope is you will see me try and, if I fail, you will hear me explain why the error occurred and then see me try again. Right up there in front of you, God, and everybody.

Lots of people write about failure and risk and the rewards of overcoming fear, especially fear of failure. As with almost everything, it's better to learn by example. So I invite you to watch me risk failure and occasionally fail. For no extra charge, you get to see something else: me getting back up, dusting myself off, and getting right back to work.

Undoubtedly, some people derive satisfaction from watching others fail. This is particularly true of some in business. I'm unwilling to change who I am and waste an opportunity to share the inherent value of failing. Snickering may be the response some choose. I'm willing to risk that.

I want you with me during the presentation. I would like for you to engage. I want to share an idea, or a story, or some cool technology, or all of the aforementioned with you. I want to convey that it's ok to fail, so I type in demos.

What I hope you'll learn, besides the cool technology I'm demonstrating, is that you're not truly free to succeed unless you're free to fail.

Chapter 32. Techganic

This chapter is about how organizations and people plant ideas. I call the process "techganic"—a word that combines "technology" and "organic" and describes the different ways in which companies grow and mature ideas in financial and capability terms. In the techganic garden, I compare the seeds to ideas and the soil to execution of these ideas. I see four categories of seeds that are sown within organizations:

1. Trampled
2. Fast-growing and then withering
3. Choked
4. Successful

This analogy was inspired by a parable recorded in the New Testament, in Luke 8:4-15:

> When a large crowd was coming together, and those from the various cities were journeying to Him, He spoke by way of a parable: 5 "The sower went out to sow his seed; and as he sowed, some fell beside the road, and it was trampled underfoot and the birds of the air ate it up. 6 Other seed fell on rocky soil, and as soon as it grew up, it withered away, because it had no moisture. 7 Other seed fell among the thorns; and the thorns grew up with it and choked it out. 8 Other seed fell into the good soil, and grew up, and produced a crop a hundred times as great." As He said these things, He would call out, "He who has ears to hear, let him hear." 9 His disciples began questioning Him as to what this parable meant. 10 And He said, "To you it has been granted to know the mysteries of the kingdom of God, but to the rest it is in parables, so that SEEING THEY MAY NOT SEE, AND HEARING THEY MAY NOT UNDERSTAND. 11 "Now the parable is this: the seed is the word of God. 12 Those beside the road are those who have heard; then the devil comes and takes away the word from their heart, so that they will not believe and be saved. 13 Those on the rocky soil are those who, when they hear, receive the word with

joy; and these have no firm root; they believe for a while, and in time of temptation fall away. 14 The seed which fell among the thorns, these are the ones who have heard, and as they go on their way they are choked with worries and riches and pleasures of this life, and bring no fruit to maturity. 15 But the seed in the good soil, these are the ones who have heard the word in an honest and good heart, and hold it fast, and bear fruit with perseverance.

The concept of being able to discern which ideas are valuable and which will not succeed is as important in business as it is in any aspect of life. Technical ideas grow techganically. There's a cycle that includes planting, nurturing, and harvest. Every good idea requires good planting (execution) to be successful.

As a manager, you have to be able to identify each type of seed and give it the appropriate type of nurture or recognize that a particular seed will never thrive and focus your attention on the fertile ones. So let's examine at each type of seed and what happens in the techganic garden.

Trampled

Luke 8:5 says, "The sower went out to sow his seed; and as he sowed, some fell beside the road, and it was trampled underfoot and the birds of the air ate it up."

Similarly, in business you'll find that some ideas fall beside the road. Some of these ideas are dead or fruitless. These trampled ideas never reach the target, rich soil in which they can thrive. People trample the ideas, and nature consumes them. They are ignored to death or taken by others. Dead ideas are a waste.

A certain percentage of trampled and confused ideas fall into the category of useless. But many of them are great ideas and would produce a rich return if only they reached the right soil. Stolen ideas fall into this category, as do ideas that the original sower never fully cultivated. Other people take these ideas and use them for their own purposes. Perhaps others improve the ideas, or perhaps they merely know how to nurture them successfully.

I wish to point out that these ideas often succeed eventually. By being aware of ideas that might get trampled, you can help find the right environment and gardener to make them thrive.

Fast Growing and Withering

Some ideas land on rocky soil with just enough dirt to get them started. But they wither as soon as the moisture and nutrients are depleted from that soil. In other words, such ideas look good in the beginning but are unsustainable. Are they good ideas? Maybe. Probably so. But they aren't planted in a good spot. They are not executed properly. The heat of the day takes them out.

Note that heat comes from the sun and sunlight is required to grow seeds into healthy plants. These ideas died of natural causes. By exposing ideas to the light of examination, you can find the ones that have potential and separate them from those that look good in the beginning but don't stand up to the heat of rigorous testing.

Choked

Other ideas land in a space filled with existing ideas. The older ideas have already been executed and are ahead of the new ideas in maturity. The existing ideas are consuming the vital resources needed for a new idea to push its way to the surface. Old ideas can choke out the new idea, thwarting its growth initially and eventually killing it.

I've heard this story repeatedly: A company starts up a new division by hiring a team and tasking them with breaking into a new market. The individuals on the team have experience in the market, so this approach makes perfect sense: Bring in people with experience to get things started relatively quickly.

The team identifies and prioritizes a list of things the company needs to do to break into the new market. At the top of the list, for example, is a better pricing strategy. This means, "We're charging too much. We need to charge less." The company doesn't like the implications of this new idea and rejects this advice in favor of old, established practices.

As a result of sticking with the old and chocking out the new, the company loses an early bid. Because the early bid is lost, the company questions the effectiveness of the team and begins reconsidering breaking into the market instead of questioning how the old ideas were allowed to choke out the new.

It's hard to see beyond the old ideas that have worked in the past. But a time always comes when the old ideas need to be weeded out so that the business can move forward with a new crop of ideas.

Successful

Some ideas grow and mature successfully. In fact, they are wildly successful and yield 100x the original investment of time and resources. I understand venture capitalists invest knowing some percentage of the companies they sponsor will fail. But the companies that succeed more than make up for the failures.

Successfully executed ideas require two general things: The ideas have to be good, and they have to be tended and nurtured successfully. There's a lot more to it than that, but you have to be willing to invest the time and assets required, or the best ideas can falter.

Recognizing the different kinds of techorganic seeds sprouting in your organization can give you a perspective that allows you to nurture worthy ideas and weed out the factors that will destine your techgarden to fail.

Chapter 33. The Playing Field Is on a Hillside

This chapter is about entropy and unfairness and how people overcome them. Entropy is part of the Second Law of Thermodynamics. Engineers know entropy as a thief that steals energy otherwise available for useful and productive work (sort of like meetings). Entropy remains constant or increases throughout a closed system. It tends towards an equilibrium, but the energy required to achieve equilibrium is lost for other purposes.

Entropy implies the arrow of time. It's the property of the universe responsible for the ice melting in your beverage (and for the room temperature decreasing slightly due to an occupant's iced beverage). It's the reason the oxygen molecules are distributed with relative uniformity in the room you currently occupy.

Entropy giveth and entropy taketh away. Sometimes entropy works for you, sometimes it works against. You probably don't mind so much that the room is filled with a fairly even distribution of oxygen, but it may bug you to no end that your beverage is warming up.

In short, entropy is not fair. And it ensures the universe is a little lopsided, as well.

I recently read Steven Pressfield's book *The War of Art*. It. Was. Awesome. *The War of Art* is about starting. Pressfield presents impressive arguments for the premise that starting is one of the hardest things to do. There's a natural resistance in us that works against us at every turn. I believe resistance is entropic. And it's not fair.

We think we do enough work to get us started and nothing happens. Why? We haven't done enough work to get started. We did enough to start us out on a level playing field—enough to keep us going at a constant velocity unless and until we are acted upon by an external unbalanced force—but that's not enough to overcome entropy. The playing field is on a hillside and not level after all.

Some people refuse to accept the effects of entropy and unfairness. These people do whatever it takes to get going and take the ball uphill all the way to the goal. One such person is Kalen Delaney, a highly respected and successful SQL Server expert. Recounting how she became a database professional, Kalen recalls that she was unemployed after having been fired. But she learned about a competitor to the company that had let her go and went to work there. Kalen took the opportunity to establish herself and became the expert she's recognized as today. Those who know and admire Kalen realize she's come a long way. She overcame resistance and countered entropy.

My point? You can overcome entropy too. Kalen did it. I did it. A bunch of the people we all admire and envy have done it. Don't let unfairness stop you from taking your career to the next level.

Chapter 34.   How Is Your Serve?

This chapter is about serving others. My current job title is Chief Servant Officer. I've held a lot of job titles that described the tasks I performed. But I've never before held a title that reflected who I am and what I aspire to. If I'm going to have a label, it might as well be something that I believe in. So when Brian Moran and I started Linchpin People, I chose my current job title with that in mind.

I'm going to serve the people I lead, the people who work for me in any capacity, and the people for whom we "do the work," as Steven Pressfield so eloquently describes it. I'm going to serve people who need any kind of help I can provide. I'm going to serve people I don't know. I'm going to serve people who choose to compete with me when they could be collaborating instead. I've adopted a motto: I Am Here To Help™

How about you? How's your serve?

Chapter 35. Volatility

This chapter is about maturity as a cure for volatility. As a leader, you have to know your own weaknesses so that you can respond to volatile situations thoughtfully and strategically.

When I think about my youth, I realize a few things: My hair's been longer, but it's never covered my red neck. I was a wild child in my younger days. I cultivated a damn-the-torpedoes attitude, and I would dare people to launch their torpedoes. At the time, I felt my responses to life's challenges were courageous, a defense of who I was and what I stood for. I saw life as a series tests, but my responses had more to do with testosterone than courage. My responses were often brash, provocative, even. Many were without thought and were therefore overreactions. Some of my responses still fall into this category.

There's a saying that God watches out for fools and drunkards, and I've been both (simultaneously at times). Looking back, I understand I was exorcising some junk from my childhood. Although my path worked for me, I do not recommend the course I took. God has used age, friends, grandchildren, and a good woman to settle me down (though not in that order). I've learned it's OK to respond to life's challenges with focus. Sometimes intensity is required. But nothing is a replacement for thinking. Strategic thinking is always a good idea.

Strategy

Strategy can have negative connotations of plotting or gaming. Strategy is an eight-letter word, but that doesn't make it twice as bad as a four-letter word. Some people confuse strategy with evil or selfish plans and equate serendipity and seat-of-the-pants thinking with virtue. Please resist the temptation to think this way.

Strategy is defined as "a carefully devised plan of action to achieve a goal, or the art of developing or carrying out such a plan." The uses you put your strategy to can be good or bad. Value judgments are separate and distinct from forethought. I know people who are serendipitously evil and people who plan strategically to do good.

Thinking strategically simply means two things: You're thinking, and you're planning. Thinking is a good thing; so is planning. Are people capable of making selfish and evil plans? Sure. And some do. But not all. Strategic thinking is a tool, and tools can be used to build or to destroy. It's usually easier to destroy than build, but that's because entropy is working with you when you destroy and against you when you attempt to build.

Responding Strategically

When leading a team, community, or any effort, one of the hardest things to learn is how to respond strategically. For me, this is especially difficult when someone I'm relating with is reacting without thinking. I believe we're most sensitive to traits in others that we dislike in ourselves. Having personally grown out of acting without thinking, I dislike seeing others react without thinking because I dislike it in myself.

Reacting without thinking when people who are in a volatile state has enormous potential to set up a negative spiral! If you've ever witnessed a bar brawl, you know what I'm talking about. A couple hotheads who won't back down collide, the challenge escalates, and before you know it everyone in the vicinity is either fighting or affected by the fight.

Is fighting always wrong? Nope. Sometimes it's precisely what needs to be done. But it's rarely the first thing that needs to be done—in contrast to the thinking of the typical hothead.

Volatility

I describe hotheads as "froggy." They're ready to jump at the least little thing. Over the years, I've learned many do this out of fear. The basic idea is to gain the upper hand in an argument because they're afraid. Afraid of what? Almost everything. Others are merely lazy, acting froggy without thinking. A minority have no fear and refuse to think. The absence of fear combined with the absence of thought is dangerous.

The propensity to act without thinking makes you volatile. Volatility, regardless of the root cause, has no place in business, especially on a technical team. I know. I was volatile, and it cost me opportunities.

The Cure

Ask yourself the question my former boss and mentor Ben McEwan used to open meetings: "What is the problem we're trying to solve?" If you can't answer this question, that's a clue that you're about to act without thinking.

The remedy for volatility? Self-awareness. Every twelve-step program starts with admitting you have a problem. Volatility is a problem. If you're volatile, admit it. The next steps are all similar to a 12-step program, too.

Without a time machine, you cannot undo the damage your volatility has caused. But you can break the patterns of behavior by thinking things through. You can cease responding without thinking. You can be less volatile every day. Growing older didn't help me nearly as much as growing up did. Time doesn't bring wisdom; it brings age. Introspection is healthy, as are counseling, mentoring, and coaching. We're all growing and changing. None of us are where we will be. Volatility is short-sighted and the antithesis of strategy. It may feel good in the moment, but it does not produce.

Chapter 36. Engines of Loss and Gain

This chapter is about winning. I like watching NASCAR races. On the surface, a race looks like a bunch of folks driving fast on a circular course. But there's much more to it than that: There's engineering and strategy, and frankly, a little luck. Most important, NASCAR is a well-run business that has built an extremely dedicated community of followers. Behind the success of NASCAR is an understanding of winning and losing, as well as an understanding of how to keep a community engaged. A NASCAR race is a lot like life when you look beneath the surface, and I've taken lessons from NASCAR on winning and losing that apply to business and communities such as the SQL Server community.

In a NASCAR race, 43 cars start and one wins. If you do the math, you realize that 42 cars do not win. The ratio of winners:losers is 1:42. A NASCAR race is an engine of loss. Why? It generates way more losers than winners.

Does that make it any less thrilling to watch? Not at all. If you look at the schedule for a year, something like 34 races are slated over a nine-month season. Frequency allows for more winners than the 1:9 ratio would suggest at first glance. Frequency helps because even though there are more losers than winners each week, there are a lot of weeks and therefore a lot of opportunities for different winners. Constraints come into play, as well. NASCAR has strict rules about the cars and various aspects of racing. Constraints on top of frequency also mitigate the winners:losers ratio, allowing the community of spectators to hope that their favorite driver has a chance to win the next race.

Unlike NASCAR, some communities fail to understand the importance of providing enough opportunities expand the potential pool of winners. Some communities don't know how to manage constraints in a way that opens the door to allow newcomers or experienced unknowns to be winners occasionally.

Community events that occur infrequently, that consistently have more losers than winners, and that have indefinite or shifting constraints damage their community. The effect of loser-laden events in such communities is like NASCAR limiting its race schedule to three races a year and declaring that only the winner of the first race is allowed to win future races.

Solution (and Anti-Solution)

Giving a community opportunities to win is not a zero-sum game. Neither is business. So what's the answer? Create events with more ways to win and allow more winners and fewer losers. Optimally, create events in which no one loses and everyone wins.

How? First, you can redefine winning. Winning has long been associated with not-losing. Some people confuse the two. In fact, it's a remarkably subtle shift to measure yourself by counting those behind you. This leads to an incredibly negative spiral that can make you believe you're winning—indeed, can only win—if others are losing. Rather than build up, such people tear down. If the goal is to win, after all, someone else must lose. Right? NO!

It's possible to create. Period. Create what? Reputation, value, good will, whatever you want to call it. You do not have to achieve, reputation, value, or good will by taking it from someone else. It can be generated.

Here's a simple test to determine the direction of your spiral: Which way are you looking? If you're measuring yourself by all the people behind you, you've fallen into this subtle trap. If you're measuring yourself by being better, helping others, learning more, improving every day, you're facing the right direction.

One way to enable more winners and create value is competition within yourself. Take online gaming for example. Online games are contests that require people to compete with each other. But success in these contests is driven by—and drives—an internal competition to reach higher levels of achievement. Competing hones skills and enables the player to reach higher levels. Achieving higher levels means the player is better equipped to compete. It's a positive spiral built on self-improvement. Some games even award points towards levels when you help someone else improve their points or level. Contrast this with a 1:43 chance of winning, and you start to get the picture of how community can spread winning among its members.

Increase the Odds

The SQL Server community is an incredibly engaged and eager-to-serve group of database professionals. Members of this community have to compete for opportunities to speak at community events, and the experienced speakers are the ones who are selected first. The draw of experience is understandable, but it limits the chances for talented newcomers to establish a reputation for themselves.

These up-and-coming experts usually get the incorrect old response: "Sorry. Try again next year!" But this defeats the purpose of community. It's the community equivalent of a No Vacancy sign outside the Old Boys Club.

It's hard to think up ways for more folks to win an opportunity, but I believe that's only because we're in the habit of creating events and opportunities where a few win. I believe this kind of thinking is so 1998.

An example of creative thinking that has provided new chances for community members to share their knowledge and grow their reputation is called SQL Saturday. In the SQL Server Community, SQL Saturday has provided opportunities for those interested to present their ideas in front of an audience eager to learn more. SQL Saturday increases the odds of winning by providing more opportunities to present. By offering local events that increase frequency, the SQL Server community is increasing the number of potential winners, just like NASCAR has done.

I think this new approach from the SQL Server community serves as proof that you can create opportunities for community that can benefit business, as well. Creating opportunities where everyone wins is a great idea. That's the principle behind the business I'm leading today.

Chapter 37. Associate of Applied Science

This chapter is about education and professional certification versus hands-on experience as job qualifications. Strictly speaking, I'm not qualified for any job that requires a Bachelor's degree because I don't have one. But I do have many years of successful experience in my field. Is a degree or certification alone really a valid way to judge a candidate's ability to do a job? Or is practical experience equally or more important?

I regularly receive email from recruiters asking if I'm available and interested in a position. Sometimes they send along the position responsibilities and qualifications, and sometimes they wait until I request more information before sending them along. I respond to most of these requests, thanking them for contacting me. I promise them I'll forward the email and the recruiter's information to anyone who is qualified and seeking a new job. Almost all of the qualifications include the phrase "Bachelor's degree required."

These requests get me to thinking about the candidates I interview for openings on my team. When I'm considering a candidate for a position I need to fill, I find myself wondering about the value of a degree or certification as opposed to solid experience. I've seen professional certifications, such as the Microsoft Certified Systems Engineer (MCSE), become inflated indicators of a candidate's potential and then completely lose all credibility because the certification required only theoretical knowledge rather than a mixture of theory and practice. I have to wonder if the same concern about experience applies to candidates who have bachelor's degrees.

Not long ago, when a technical professional achieved the MCSE certification it meant a lot more than it currently means. The certification went through an amazing cycle: First, there were very few MCSEs. Only the most passionate and experienced IT pros attempted the certification exam. Those that were certified did great work, word got out, and certified people became highly respected and were in demand. This meant the certification was also in high demand. Training companies picked up on this trend and started MCSE boot camps that not only promised you could pass the tests after a week or two of their training, but they guaranteed it.

The result? Many MCSEs were on the market, but many of them lacked practical experience to back up the certification. Once again, word got out, and the value of the MCSE certification decreased.

Bachelor's degrees used to be sparse in the candidate jungle and meant the student had achieved something relatively few could achieve. A minority of people entering the job market had Bachelor's degrees when I entered the job market out of high school back in 1981. Nowadays, it seems almost everyone has one. Maybe that's why they're required to get most jobs.

But does having a degree mean the candidate really has the ability to understand and perform the day-to-day duties of a technical job? In theory, a person with a degree has shown the ability to learn and adapt, which makes them desirable employees. These qualities are great if you have plenty of time for training the person and helping them develop the skills they need to actually perform the job.

If I need an employee who can ramp up fast, I want someone who has experience. A degree or certification is a plus, but it's not the one thing I'm looking for. As a hiring manager, I'm beginning to wonder if the Bachelor's degree is in danger of going the same route as the MCSE certification. Is it at risk of losing value in the marketplace as it becomes more common and has no practical component?

Finding an ideal new hire is an art. The person has to be a cultural fit in addition to having the job skills and ability to grow. I don't want to dismiss the value of education and certification, but I can't emphasize enough how important balance is in this area as in all others.

Chapter 38. Freedom to Innovate

This chapter is about the removing obstacles to employees' performance, and it was inspired by two sources: conversations with my business partner Brian Moran and a comment in an anecdote from The Art of Possibility, recommended by my friend and brother K. Brian Kelley.

Brian Moran and I are intentionally and deliberately defining a philosophy for our new business. We believe it is important and worth the effort.

The comment from The Art of Possibility is from an anecdote on pages 37 and 38 of the paperback edition. It's about an orchestra conductor's response to a musician who was disengaged from the piece of music at hand. After the conductor found a way to engage the musician (he *listened* to her complaint – crazy idea), he had this comment: *"When I had been viewing her as an unimportant casualty, I had to pretend it did not matter that for some reason she was not engaged. Meanwhile, I wasted energy both watching and ignoring her."*

Wow. Think of all the energy wasted in companies today. Think of all the energy wasted in communities viewing members as unimportant casualties (vocal minorities).

The conductor's final conclusion after communicating with this individual: "The lesson I learned is that *the player who looks least engaged may be the most committed member of the group.*"

I think this is an important lesson for all managers. We need to listen to team members and acknowledge that they want to perform well. Our job is to remove obstacles and make sure employees can contribute freely and feel that they personally gain from doing so. One way we can do this is by changing our thinking that there's a difference between what benefits the company and what benefits the workers.

Blocking Requests

I believe people want to perform well—in life, on the job, in business, and in personal relationships. I think wanting to do well is the natural state of most people. People are basically good. There are some exceptions, obviously, and I think society has mechanisms for correcting deviations. And all people deviate from time to time through habit, error, or poor choices. But essentially, people want to do good and be good.

I think leaders often get in the way of people's inherent motivation to do good. Most of the getting in the way is unintentional. (Sometimes it isn't unintentional. However, some people derive pleasure from manipulating others.)

Leadership makes requests that are often reasonable and must simply be executed. This is the use case for military orders delivered during combat. But most of us aren't in physical combat. (God bless those who are. I pray for them.) The closest we come to a military-style must-execute scenario is agile business or development environments where fast action and response is critical to success. The leader may have access to more information than is available to those she leads.

In these scenarios, communication is vital. But even more vital is a history of trust and respect. If the leader is not the type of person to abuse authority merely to enjoy the reactions of others, the leader is more apt to obtain the trust and respect of those they lead. Professionals want to know why they're being asked to do something because they want to do a good job. If they know *why*, they will be able to adapt to and overcome obstacles – perhaps reducing the time needed to accomplish the task or adding success to the achievement.

Holding up a hoop and expecting others to jump through, ordering "Jump!" to hear others ask how high—these are simply blocking requests.

What do I mean by blocking requests? I mean requests that serve no real purpose other than to boost the ego of the person issuing the request. The type of leader who issues blocking requests is simply keeping the recipients from using their brains to do something meaningful.

The human brain is an awesome processor. I recently read an estimate that the total computational power on the planet at this time is comparable to that of the computing power of a single human brain. While that represents the astounding amount of capacity required by conscious thought, it reflects the fact that the human mind's abilities are also finite. People can (and do) use their minds to create amazing inventions and philosophies, to conceive art and discover science, but the mind cannot produce more mind. Conscious attention is a subset of the working cognitive brain power available. Blocking requests—holding up hoops or ordering others to jump—consume brain cycles better spent on other tasks.

An Example

In the current market, talented technologists move from venture to venture every couple years. While they're at Venture A, they learn stuff that inevitably finds its way into their work at Venture B; stuff that inevitably finds its way into their work at Venture C. And so the cycle repeats.

Companies see this movement as creating an opportunity for developers to take intellectual property with them to their next job. Trying to protect their intellectual property, these companies pay lawyers to draft long contractual documents to ostensibly prevent industrial espionage. But these contracts can also stifle competition and innovation, as in the case of the gifted technologist I knew. Plus, non-compete agreements fail almost every time they come to trial because many of the cases come down to an individual's ability to earn a living.

Let me tell you about that gifted technologist I just mentioned. He is well known in his field as someone who's created elegant solutions to complex problems. He's currently employed by a company that believes anything he invents during his tenure with the company—whether on the clock or off—belongs to the company. It gets worse: As a condition of employment, he was required to sign a document to this effect. The result?

He's stopped innovating outside of work. This has impacted his performance on the job, as well as his opinion of the company he works for. He longs for trust and respect. He wants to love his job. But this agreement effectively limits the amount of effort he desires to put into work outside of company employment, which in turn undercuts his continuing education, which in turn limits his effectiveness at work, which in turn reduces his performance, which in turn depresses him. It's an engine of loss.

My friend wanted to do good work and be a good employee. The company wanted to protect its interests and the possibility of fraudulent exploitation of its resources. All the intentions were good, but the result was not. What was left out of the relationship is human nature and the need to be creative. People grow. Technologists learn stuff. They play. It's in their nature to do so. But non-compete agreements stifle their creativity. And (aside from the fact that such contracts are not enforceable) these agreements ultimately, present a barrier to excellence.

One Solution

What would happen if businesses adopted a solution that recognized the conditions of modern economics for gifted technologists? What if, instead of punishing the technologist for learning as she grows from engagement to engagement, business practices accepted or even encouraged the nomadic nature of the technology market and the good will inherent in most employees' desire to be creative? What if we stopped lying? What would happen if we embraced the truth?

What would this look like in practice? For one, there would be *no more non-compete agreements*. These innovation-killing non-compete agreements would go the way of the horse and buggy—quaint, nostalgic, but not useful in modern practice.

Another change would be *ownership agreements* in the place of non-competes. Imagine ownership agreements that express that the ~~serf~~ inventor of an idea owns the idea. (I know! Crazy, isn't it?) This would extend to works derived from the current employer's intellectual property. Business leaders and MBAs are of two minds when it comes to derivative works. On one hand, they don't mind it one bit when they derive something from another patented process and profit from it. On the other hand, they expressly forbid as many others as possible from doing the same thing to their inventions.

This could create a win-win for the inventor and the company by encouraging employees to innovate and do good work on the job. Once technologists are free to innovate, they can produce unimaginably awesome technology. And again, most technologists are good people who want to do a good job, so they're more than willing to share the spoils with the employer who allowed (or *facilitated*, even!) their success in innovation. (Note: I'm speaking of a minority stake.) This is due in large part to the thrill of producing experienced by every technologist I know. Once this thrill is experienced, the technologist wants to do it again. And again.

Should the business own a (minority) stake in the derivative work? Once the non-compete thinking is abolished, such a suggestion becomes plausible. It's amazing the possibilities that emerge once the threat of starving and/or bankruptcy are removed from the equation.

We Can All Benefit

If we stop thinking that one side's gain is the other side's loss, we have an opportunity to let everyone win.

Chapter 39. Love Your Opponents

This chapter is about opponents. Opposition is a clue that you need to pursue what you're working on. Don't believe me? Most worthwhile endeavors have met opposition, especially early on. It happens in life, it happens in business, and it even happens in communities. For example, take a SQL Server community series of events called SQL Saturday. This series was a grassroots initiative designed to give people an opportunity to improve their professional qualifications, both as database professionals and as speakers. SQL Saturday has now gone past its 100th event. However, something as cool as SQL Saturday was initially opposed by a major community entity that regarded the idea as competition to its own community events (without apology, to date).

What does opposition mean? The success of SQL Saturday illustrates that opposition means you're on to something. Perhaps you're beating someone to the punch, or you're delivering with more agility or quality (or *both*!). You can be certain of one thing: You have an advantage. If you didn't, why would anyone waste time opposing you? If your idea or product or service stinks, rather than oppose you, others will find it's much more efficient to observe as you wither and eventually die on the vine.

Only those who *care* will oppose. People who care fall into roughly two categories: those who have a vested interest in your success, and those who have a vested interest in your failure. Your family usually falls into the first category as they may depend on you and your success. Plus, your family usually loves and supports you unconditionally. Competitors fall into the second category as they may think that your success will reduce their ability to succeed.

Why You Should Love Competitors

Although opponents may want you to fail, they can provide a valuable service to you. Competitors will tell you – for *free* – where your idea, product, or service is vulnerable. Don't gloss over the "free" part. That's not just Andy being coy; it's a fact. Your competition may not come right and say, "Your weakness is thus and so." (At least they won't say that directly to you.) But they will tell people who are searching for your product or service.

Is pointing out a competitor's weaknesses to potential customers a bad thing? I suppose it depends on how the person doing the pointing wishes their own customers to view them. Customers will perceive the negative person as a big fat hairy loser. You see, people are smart. They understand when they hear someone speaking badly about anyone –competition included. People understand when someone is communicating their own capabilities by characterizing others negatively.

Each of us understands that if somebody talks negatively about someone else, the negative person will say bad things about us.

Do you really want to communicate to your customers (of all people) that you're capable of talking negatively about them? Me neither. What's the alternative? Trust and respect. You can even make money from trust and respect if you engage in coopetition instead of competition. Try it. It works.

Of course success is the greatest counter to opposition, as SQL Saturday illustrates. But you can also listen to free advice from your opponents, acknowledge it, and grow from it. Your potential customers will respect you for your transparency. If you make transparency the rule that guides your personal and professional behavior, the way you relate to those in the same business becomes increasingly important. Your best defense against opposition is to be professional, play fair, and treat others as you would like to be treated. (I've read that somewhere else.)

Chapter 40. Outlasting Outrageous Opposition

This chapter is about responding to outrageous opposition. Opposition happens. As I stated in the previous chapter, opposition is a clue that you need to pursue your goal. Does that mean all opposition is good or can be put to good use? Nope. Some of it is troll-ish noise. How can you tell the difference between someone who is competing with you and someone who is capable of dangerous, destructive behavior and actions against you or your enterprise? Hindsight is unfortunately the only definite answer.

I wish there were better ways to discern between those who merely oppose your art or work and those who will purposefully engage in activities designed to impede or destroy it. But I've learned to recognize a few hints. I've noticed the following characteristics of individuals who behave destructively:

> 1. You are probably not their first target. There's a good chance an individual bent on destructive activities has behaved this way in the past. In an age of ever-increasing visibility, these folks leave a trail.
> 2. Lord knows what motivates them to behave in this manner, but negative people are, thankfully, a tiny minority. This makes identifying them easier, especially in social media circles. Start by identifying the beginning of destructive behavior and then see who becomes emotionally charged. Emotional outbursts help you identify potential culprits. When you see negativity repeated and spread across several incidents, you can narrow down a pattern. In many social media communities, two incidents is often enough to isolate an individual attempting negative activity.
> 3. Destructive efforts usually come to naught. Visibility of the negative person's efforts is their greatest ally and their Achilles' heel. If they become too visible, they're exposed for what they are. If they're not visible enough, no one notices. Destructive behavior

is parasite of social media communities, but the parasite can kill off potential hosts when people recognize the parasitic behavior before more serious harm can be done.

So how do you respond effectively to negativity trolls? Ignore them. Again, you don't know their motivation, but anyone who's been around a five-year-old who isn't getting her way understands that the underlying goal is attention. If the negative people don't get attention from you, they will move on.

Chapter 41. Ensuring that a Metric is Effective

A local fast-food establishment measures the amount of time between drive-through order placement and order delivery. Their goal is to minimize this time and create a fast-moving drive-through experience for patrons. That's a great goal for everyone. The restaurant wins; the customer wins.

Sometimes the drive-through staff games the system. The time from order to delivery rocks, but they accomplish this by queuing those waiting to place their orders. This puts one car and presumably one order in the queue between order placement and delivery. The crew of the establishment is able to focus on one order at a time, so their measured delivery times are exceptional. The problem with this is the metric no longer makes for a positive customer experience at the drive-through.

There are lots of options when it comes to solutions. The establishment could change the metric, eliminate the metric, or fire the employees who are responsible for gaming the system. Let's look at each of these options.

Change the Metric

The metric can be changed if it doesn't accomplish the goal. In this example, the goal is to provide feedback to give drive-through crew a way to understand their effectiveness with customer experience. When the crew doesn't game the system, the metric produces the desired result.

What motivates the crew to game the system? They want to demonstrate success against the metric. There could be some reward attached to achieving the best numbers. It may not be a monetary reward.

As a side note, a lot of studies show money isn't the best reward for certain types of behavior. *But* these studies are often misinterpreted to justify not paying people enough or not giving people raises – even cost-of-living raises. The lie is: Money demotivates. The truth is rather simple, so simple it's very easy to overlook. The truth is that love motivates. The response should never be: Don't pay people more. The response should always be: Don't kill the love. Remember, the crew wants to succeed.

Why would someone kill the love? I will answer that later, but I will offer this teaser: What do Julius Caesar, Napoleon Bonaparte, and your average freshly minted MBA have in common? The answer involves control.

Eliminate the Metric

If the metric is encouraging the absolutely incorrect behavior, you can simply ditch the metric. On the surface, this seems extreme. But let's think about it for a minute. Most metrics are designed to provide feedback during the process. The logic is sound: The earlier you discover a problem, the sooner you can correct it. The sooner you correct a problem, the less expensive the problem is to solve.

This reminds me of an analogy about asteroids that I use for software projects. Theoretically, there's a point where you can deflect an asteroid heading towards Earth with a very small amount of force. Later, when the asteroid is closer to Earth, all the nukes on the planet won't help.

But if the metric isn't doing the job because it's being circumvented, it might just be better to move on and find a new means of measuring effectiveness. In this case, abusing the metric by gaming the system has produced the opposite of the intended effect at the local fast-food establishment. It's moved the *problem* to earlier in the process by queuing patrons before they place their order, thereby magnifying the problem (from the customer perspective).

How is the problem magnified? For one thing, the fast food team has generated less capacity for cars at the back of the line because cars are not queued between the drive-through speaker and the delivery window. A key tenet of the Theory of Constraints states, "Losses accumulate. Gains don't." By starting the losses earlier (and by design), the team has created more opportunity for loss-accumulation.

Fire Those Responsible for Gaming the System

Let's seriously consider the third option, firing the people who are responsible for undermining the metric. Who *is* responsible for gaming the system here? You can make the argument that it's the crew. They were presented with a measurement and circumvented the intent of that measurement. You can call that wrong or even evil. Or you can call it clever. The survival instinct is in play. People are intrinsically motivated to do a good job, and the team has adapted the metric so that they are meeting the metric they were asked to meet.

You can also make the argument that the person in leadership who chose the metric is responsible. That person picked a metric that could be gamed. Or the person added motivation to the metric that over stimulated the workforce at the fast-food establishment.

As a leader who has been in situations that required creating a metric for success, I believe the leadership is more responsible for this outcome than anyone else. Had it worked, the leader would certainly take responsibility—you can count on that.

A Better Solution

The real problem is not that the metric is bad. The real problem is that the employees had to get creative in order to meet the metric. Maybe it was over-enforced and given more credence—more weight—than it deserved. The employees responded out of self-preservation. This imbalance motivated gaming the system.

I would argue that there's a single metric that counts in the real world: shipping. Did you deliver? Does it work? Is your customer delighted? These are simple yes or no questions, and at the end of a project they are *all* that matters. Interim measurements are good and useful as long as they serve their intended purpose. When they are misapplied, expect gaming.

Chapter 42.   Evil Is Easy. Creating Is Hard

This chapter is about contrasting destruction with creation. Destruction holds back business, community, and progress—usually way more than anyone intends. Creating drives business, community, and progress.

In *Star Trek II: The Wrath of Khan*, Mr. Spock said, "As a matter of cosmic history, it has always been easier to destroy than to create." Destruction, or evil, can get away from you fast. Before you know it, you've done more than you intended, or wanted. You've hurt people you never intended to hurt. And while justifications like collateral damage may temporarily placate, they end up sounding hollow. Sometimes for years.

It takes very little imagination to react to a wrong, to say, "I told you so," to a team member or subordinate, or to feel smug when your competition fails. It's natural, in fact. Because it's so easy, evil is lazy.

In contrast, creating is hard. Building something from nothing is difficult, to say the least. But often it's even more difficult to *conceive* an idea worth building in the first place. I have to admit, my best ideas have come to me in dreams. I'm blessed to have a business partner, Brian Moran, who comes up with great ideas every day. I like implementing ideas, so we make a good team.

The main reason creating and conceiving ideas is difficult is that thinking is hard. I don't know how you think. I know how I think, though. Maybe you think similarly; maybe you don't. I actually set aside time each week to ponder. It's a cool time most weeks.

If thinking is so difficult, why do it? Sometimes I come up with a killer idea (like SQLPeople! Which started as a dream). Thinking and creating is so worth it when you get to share that idea. It doesn't matter if you get to share it with someone in a one-on-one mentoring session or in front of a user group or presenting at a community event—it is awesome to share!

And this is why I love community. It's more than giving back. Giving back sounds like you're paying off some old debt, and that's *not* how contributing to community feels at all. Giving back is a transaction that flows from one person to another or many. That's not community! Sharing is community. Community is engaging. Community is participating and creating.

Community is everything that destruction is not. It requires the hard work of creating, but it counters destruction. One reason why I started writing this book is to encourage people to create, and to discourage those who would block creativity and innovation. Creating is hard work, but it is *so* worth it.

Chapter 43. Performance-based Management Doesn't Work

In *Mere Christianity*, C. S. Lewis refutes an argument by countering, "It has every amiable quality except that of being useful." I feel that way about performance-based management (PBM). I am a metrics person. I thrive—intellectually, emotionally, and economically—on business intelligence, and Key Performance Indicators (KPIs), and dashboards. I love data mining and predictive analytics. Measurement and analysis appeal to my engineer's nature and "instrumentation-eer's" heart. So when it comes to PBM, you'd expect me to be all in. And I was, almost.

There I was, sitting in the cat-bird seat. We were a team of five, charged with expanding a successful data warehouse. We had a person who wrote special one-off applications for data mining, an awesome business analyst, a great report-writer, a guy who knew the source system like the back of his hand, and me, the SQL Server database guy.

The company had implemented a 20-60-20 PBM scheme after someone whose title began with the letter C read a book and thought: "Why are we wasting all this time *thinking* and *leading* and *managing* when we could just lump people into one of three buckets and be done with it! Think of all the time we could spend reading more cool books!"

(Ok, I'm not sure why it was implemented; that's just my theory.)

With five people on the team, the PBM math worked out perfectly. We would have one top-20 person, one bottom-20 person, and three middle-60 people. Awesome. Except how would we determine who belonged in which category?

By luck of the draw, I happened to solve the big-problem-du-jour the week before the managers were to submit their suggestions for rankings. I won the PBM lottery, as it were. The person who had been in my position previously, and who had contributed to my success substantially, was ranked last. The other three were lumped into the middle 60.

Here's my first question: If we are a team, we each have vastly different roles, and we are each good at our job, how do you determine who outperforms the others? PBM has a smarmy answer for this scenario, and that answer has every amiable quality except that of being useful.

What really happens in this scenario? It turns out that it takes a village to build and maintain a successful and useful data warehouse project. In other words, a team. When everyone contributes to the success of the project, a positive spiral is created. Everyone realizes we are our brother's and sister's keeper, that the success of all hangs on the success of each. What's more, teamwork is *easier*. It requires real effort to produce anything of value single-handed.

But wait, there's more! Because more eyes are on the work, quality improves. The quality percentage for a useless data warehouse—the ratio of good data to bad data—is a surprisingly high number. This is due to how the data is used, mostly in aggregation. Constraint Theory teaches us that losses accumulate, gains don't. In a data warehouse project, the impact of incorrect data or the incorrect application of data is exponential. If you don't believe this before your first data warehouse project, I bet you will afterwards.

It turns out PBM's "friendly competition" kills teamwork faster than anything else. Good people feel less motivated to help because they are punished for the success of others.

In the version of PBM that our team suffered through, the top 20 person got everything he or she wanted. The bottom 20 person was basically ignored until he or she quit or was fired. The middle 60 were alternately tolerated and encouraged to be more like the top 20 person.

But we were all good at our jobs!

That didn't matter. Only the buckets mattered.

And this is one of the reasons PBM doesn't work: It kills teamwork.

My Question

Upon learning the mechanics of PBM, I asked the following question: "Are we hiring the wrong people, or are we mismanaging the right people we hire 80 percent of the time?"

I think that is the right question. But I didn't get a good answer.

I have witnessed many peers subjected to PBM. Everyone suffers. PBM is an application of the manufacturing mindset to modern business, and it fails to recognize important facets of creating technology and business success. The goal of PBM is equally noble and unachievable. I think there are better ways of managing and measuring a team if you've hired the right people and are getting the job done.

Chapter 44. Data Visualization and Dashboards

This chapter is about the need for data visualization and dashboards. Dashboards are where data meets decision-makers. The field of data visualization is about this intersection of information and actors. A dashboard translates the numbers and communicates their significance in a manner that is clear enough to define action. Decisions are supported by data visualization systems. Business and intelligence meet. The data is right there, represented numerically, or graphically or both, waiting to be used.

I'm an advocate for data visualization and dashboards, but one aspect of data visualization that you have to keep in mind is integrity—both your integrity and that of the information. Data visualization specialists, like Edward Tufte (one of the most prominent names if data visualization and information design), value integrity in communications. (At least, the good experts value it.) Tufte has earned a reputation for clarity and insight, and I recommend reading his work if you're interested in this topic. Let me tell you a couple of stories from my past experience that explain what I think about integrity and data visualization and dashboards.

Simple Is Good

Dashboards are elegant. They do not have to be complex. In fact, the most effective data visualizations are intuitive and almost instantly convey the desired information. Simple is good. I learned this lesson in one of my previous jobs.

A long time ago (back when the years began with the number 1) in a place far, far away, I built a Manufacturing Execution System (MES) called Plant-Wide Webs (PWWs—catchy, eh?). It was one of the first MESs to exclusively use a browser for visualization. The idea of PWWs was to convey – at a glance – the state of a manufacturing facility or enterprise. Data visualization.

As I mentioned in the chapter on performance-based management, the only metrics that count are shipping and delighting customers. I believe that's true for measuring employee performance, but I also believe it's true for anything you want to measure. Behind that statement is a principle, and it is this: I believe it is possible to isolate or create an effective, single, accurate-enough metric for anything. Is the metric going to communicate everything that's going on at all levels of your business at a glance? Goodness no. But, I maintain it's possible to glean way more than 80 percent of the important truth from a single number, and that's what PWWs did.

The PWWs dashboard? It was a modified stoplight:

I started with an electronic drawing of the facility, which I converted to an HTML image map. The map was completely green, yellow, or red. The numbers behind this were not simple, but they were available (via a click or two, maximum) and condensed into a single metric. And they were near-real-time and immediately recognizable. Back in the day, "near-real-time" meant accurate to about a minute. The plant manager could view the facility's performance in near-real-time all day.

History was provided (of course), and drill-through was supported as well. After all, drilling was as simple as linking; something at which HTML excels. Each click would take the manager to more detail. The first click on the plant image would be a copy of the image split into several shapes, each representing a section of the plant and each reflecting the red, yellow, or green status of respective facility sections. And each section was drillable. And so on, until you reached a screen filled with readings from actual machines – data collected from data acquisition systems or Programmable Logic Controllers (PLCs) or Human-Machine Interface systems (HMIs).

The graphics then (and certainly not today) were not earth-shattering. But visualization was easy to maintain and scale, it was fast-rendering, and best of all it was simple and clear. It *communicated*. What did it communicate? The state of the plant? No. The state of the *business!*

Data Visualization Can Be Dangerous

Users aren't stupid. They are your community. If you treat your community like they are stupid, you make more work for yourself. You also communicate that you distrust and disrespect them. Transparency isn't merely the right thing to do, it is also the smart thing to do. I learned that lesson when I naively created a dashboard that disclosed facts that my boss wasn't so sure should be disclosed.

On my first data professional gig, I was hired to implement and manage the reporting solution. Back in those days I could hold my own as a web developer. The short version of a long story is: I had fixed some ported code, when the manager decided to move the current SQL Server person to another position. Since I was the only other person with the words "SQL" and "Server" near each other on my resume, I got the gig.

Now mind you, I *thought* I could do the job. When my manager asked me if I could do it, I told him, "Yes. How hard can it be to tell developers, 'No'?" <CaptainSnarkyIWas>. I learned a lot during that first real database person position.

I did a few things right, though. One of them was to trust my community. We started with a pilot of ten power users. They were all internal, part of our same company. But the next step was to expand to something like eighty users, and not all of them worked for us. So they didn't have access to all the information available to the original group.

Expanding a pilot to include more people sounds like a simple thing. Here's why it wasn't:

In ETL operations there is this thing called latency (an engineering idea) that is tightly coupled and indirectly proportional to another thing called throughput. The more stuff you can shove through a pipe, the less latency you experience. Back then, we were loading a ton of data, relatively speaking. It took days to load a couple tables in our data warehouse.

Since we didn't want to wait around the clock, I found an old spreadsheet I'd created to do predictive analytics. My idea was to sample the current number of rows in the destination table every now and then, along with the time. And then we'd do some math, and then do some more math. And then we'd have a science-backed wild guess about when the table would contain all the rows from the source.

That inspired another idea: I could build a web page to display the latency metadata used as the source in part of the calculations. It was a fantastic internal tool. That inspired another idea: Why keep it internal? Most of the calls I was fielding were from users outside the company who had no idea why their report numbers were changing with each refresh. Everyone *inside* knew when there was an issue with the overnight batch processing that increased latency. But not those outside. So I placed a link to the latency page on the website and published the latency data with our next release.

I almost got fired.

My boss considered that latency data proprietary because it basically showed when we were not compliant with our SLA. I get that now. (I didn't back then, but I do now.) It never occurred to me that we should withhold information the users needed if they were to make informed decisions about the validity of the information we provided.

Lack of transparency in such matters still doesn't occur to me. Transparency stopped my phone from ringing with complaints about the system, allowing me to concentrate on more pressing (and valuable) matters.

I probably would have been fired if the bosses of the external users hadn't called my boss's boss to tell her what an awesome idea that was. Just about the time my boss was ready to chew me out for releasing proprietary performance and SLA data, he got a call from his boss. She chewed him out for not letting her know about this cool new initiative that was saving our customers time and increasing the value of our data and service to them.

I trusted my users, and it was the right thing to do. If a dashboard is valuable to your users, it's the right thing to do. But maybe it's also a good idea to be aware of how others in the organization will react to sharing information.

Elegant! Equals Pretty

Data visualization and dashboards are about conveying information in a compelling visual way. A dashboard needs to look good, as well as being functional.

I tell every student that attends my SQL Server Integration Services (SSIS) class, "Anyone can build SSIS packages that work. I expect your SSIS packages to also be pretty." But I leave them with this caveat, "If you have to choose between pretty and functional, always choose functional."

The same goes for dashboards. If you are afforded the time to delight the customer, do so. If not, opt for "working" over "pretty" every time. Make it as pretty and fast as you possibly can, right after you get it working.

Remember one of my favorite sayings: If you deliver quality late, no one will remember it's late. If you deliver junk on time, no one will forget it was junk.

I've never had a customer or user come back to me after I've delivered quality late and say, "Sure, Andy. This works well and all, but you were two days late." Customers simply do not remember that a project was late if it does what they want. But come in early and under budget with bugs? You will not hear the end of it.

Communicate

A dashboard is simply a communication medium. It translates data into actionable information. It's that simple. If your dashboard does amazing things but sacrifices any portion of this vital function, then your dashboard stinks. Get this part right. Communicate the state of the business with integrity, quickly, and accurately. Provide multiple levels (grains) of information. Trust and respect your community for they can make your job easier or more difficult.

Chapter 45. Credibility

This chapter is about credibility. I see credibility as consistency between values and actions. When what I say matches what I do, people will judge me credible. When what I say does *not* align with what I do, people will subtract credibility from their estimation.

I believe everyone has an Internal List of Acceptable Actions (ILAA). I read this in a book about values-based leadership. Before I read about the ILAA, I knew it existed, but I didn't have a name for it. I believe the ILAA is a sorted list and that the sorting is in order of most-acceptable to least-acceptable. Acceptable to whom? Others, community, or society.

ILAAs share some characteristics with and are a manifestation of our individual consciences. The rules dictated by the human conscience are comparable to Isaac Asimov's Three Laws of Robotics: "A robot may not injure a human being or, through inaction, allow a human being to come to harm; a robot must obey the orders given to it by human beings, except where such orders would conflict with the First Law; a robot must protect its own existence as long as such protection does not conflict with the First or Second Law."

Actions that a human may or may not take are coupled to—or are instances of, if you prefer—the rules embedded in our conscience. For example, rescuing a kitten from a life-threatening situation should be near the top of everyone's ILAA.

Of course, humans are not robots. Our consciences differ from one human to the next. As a result, our individual ILAAs vary. In most cases, the variance is small and may be insignificant. In some cases, the variance is extreme. In my opinion, ILAAs are the best metric to determine the values of others. At a minimum, observing the actions of others reveals what they considerable acceptable.

Credibility and Values

Values are exposed when we observe how people act. I call this *listening to what people do*. An interesting observation I've made is that a person's ILAA and values can be unacceptable to most human beings, but such a person may still have credibility. Credibility sounds noble. But I consider credibility neutral. I believe a person can be credible and yet hold values detrimental to society, other individuals, or a community. Some people that I consider credible hold values that I simply disagree with, but they are consistent in their communication and action. Therefore these people are credible.

I believe once a person has demonstrated he or she is capable of unacceptable behavior—to anyone, for any reason—that person has demonstrated he or she is capable of that same behavior towards me. Why? It's on their ILAA. They just proved it through their actions towards others.

If you witness a coworker take something that belongs to another coworker, the person doing the taking is demonstrating he or she is capable of stealing from you. If you observe your business partner treat another partner, or a competitor, or a customer, or anyone unfairly, your partner is communicating he or she is capable of treating you unfairly. If you see an organization abuse one person, that organization is demonstrating that you could be next. If bad behavior is on someone's ILAA, everyone is a potential target.

Although circumstances may be used to excuse, reason, or justify, circumstances do not apply here. ILAAs are about capability. Remember, this is about what is inside the individual.

Interpretation

Of course, it would be awesome if credibility were as simple as consistency of action, but it's not. There are at least two areas that are subject to interpretation: the interpretation by others of what I say, and the interpretation by others of what I do.

What I Say Is Important

What we say says something about what's inside. In the Court of Public Opinion, everything I say is used for or against me. That's normal and fair and isn't going to change. For these reasons, it's important to consider what I say (or write). Things are always lost in translation, regardless of the communication medium. Email is particularly susceptible to misinterpretation.

Even if I'm very precise (which I am not), what I say and write is subject to interpretation. This is *especially* true about what I write: I've been told I am "direct" in writing. That likely stems from my training as an engineer, but that is merely an excuse. It does not help someone reading my "direct" email to think, "Maybe Andy is just being an engineer." And, my directness leads to confusion in the instances when I am sincerely communicating something unpleasant in a direct manner. (It happens.)

I have learned I never have to explain or apologize for things I do not say or write. This has made me less likely to contribute to conversations when I don't have something positive to add.

In matters I disagree with, I find Gamaliel's strategy appealing. (See the quote above from Acts 5:34-39.) After Jesus' death, His followers were still stirring people up. The rulers in Jerusalem had the followers arrested and then told them to stop. But the followers refused, which infuriated the rulers. As the rulers considered what to do next, Gamaliel spoke:

> ...a Pharisee named Gamaliel, a teacher of the law, who was honored by all the people, stood up in the Sanhedrin and ordered that the men be put outside for a little while. Then he addressed the Sanhedrin: "Men of Israel, consider carefully what you intend to do to these men. Some time ago Theudas appeared, claiming to be somebody, and about four hundred men rallied to him. He was killed, all his followers were dispersed, and it all came to nothing. After him, Judas the Galilean appeared in the days of the census and led a band of people in revolt. He too was killed, and all his followers were scattered. Therefore, in the present case I advise you: Leave these men alone! Let them go! For if their purpose or activity is of human origin, it will fail. But if it is from God, you will not be able to stop these men; you will only find yourselves fighting against God." Acts 5:34-39 NIV

How do I apply this? I believe much in life (and society and community) is organic. I believe things grow and perish organically, and that this is part of a natural cycle. To politely apply an organic metaphor, fertilizer will promote growth, but too much fertilizer will kill. Mixing different types of fertilizer can be harmful or deadly to the very things we're trying to grow, while the correct mixture will support maximum growth.

If I can help, I will. If I cannot help, I'm keeping my fertilizer to myself. If there's a problem, it will self-correct (one way or the other). If the thing is meant to be, it will be. If not, it will fail. All on its own.

I have reached the following conclusion: Sometimes what I intend for help merely distracts from the real issue and thereby prolongs the inevitable organic result.

What I Do Is Important

I noticed a pattern in my behavior this past year: I was doing a lot of things on autopilot. Most of the time this was fine. I have mostly good habits that have served me well over the years. Habits like treating others better than I treat myself, decent priorities, serving people as they crossed my path. That sounds good, but those habits, like Three Laws of Robotics mentioned earlier, are subject to failure. My habits led me to places I didn't like at times this past year.

One example is something I had to apologize for in a public way. Another example is my priorities. I had the best of intentions but my priorities were out of whack. Looking back, I now realize I had picked the best and most noble paving stones from the center of the road to Hell (which is paved with good intentions). Through engaging in a Bible study group with some friends and brothers, I believe God revealed this subtle and tragic error in my priorities. Through this same Bible study, circumstances, and prayer, God has been correcting my trajectory and continues to correct it.

One result? My word for this year is "intentional." Good habits are good to have, but they are no substitute for thinking. I have been lazy, falling back on good habits when I should have been actively engaged, thinking, and sometimes changing. I could write for hours (literally) about the stuff that's already changed in my life as a result of this reevaluation and active realignment of my priorities but this chapter is long enough! Some major areas already impacted include communication, physical fitness, matters of faith, finances, and giving (time, money, technical help, social awareness). There is more to come. Being intentional is one goal for the year.

I have learned this past year the importance of forgiveness, both giving and receiving it. Communities are a social ecology as much as a social economy. If my apology taught me nothing else, it demonstrated that our technical community is a forgiving one. Other communities in which I participate share this characteristic.

What all this has taught me is that my values and actions needed to be in sync if I'm to be credible. For me, maintaining credibility is a natural result of choosing to live more transparently. That applies to all areas of my life: family, faith, business, and everything else. I believe credibility works the same for organizations and communities as it does for individuals. I will continue to urge the organizations, communities, and individuals I love to practice transparency, and thereby garner more credibility.

Chapter 46. To Snark or not to Snark

This chapter is about communication. I enjoy listening to a good comedian and reading the works of humorous writers. Life is too short to waste on misery, and a hearty laugh is good for the soul. Some humor is educational, thought-provoking, and surprising. Some humor, though, can be hurtful.

Some jokes and comments build people up while others tear them down. I understand motivational theories that support "inspiring" people by offending them (you know, so they'll remember). But many popular and seemingly intuitive management practices based on that idea simply don't work, as evidenced in the book *Drive* by Daniel Pink. While short-term gains are possible, this sort of motivation poisons long-term productivity.

That's not the worst of it. Cleverness, or rather, feeling that you're being clever, is addictive. Some people get a thrill out of combining (real or imagined) advantage and snarkiness to achieve a zinger.

The thrill and associated endorphins are understandably pleasant—for the one being snarky. But as easy as negativity is, it's not a sustainable way to create something positive. As Spock said in Star *Trek II: The Wrath of Khan*, "As a matter of cosmic history, it has always been easier to destroy, than to create."

This holds in communications, as well. It is much easier to destroy than create. I am going to take that statement one step further: It is *lazier* to destroy than create. Inspiring people without tearing them down is hard. It is way harder than simply throwing some snarky comment in their general direction. It involves something that cannot be manufactured—your engagement.

Engagement requires attention. Web marketing people will tell you web advertisements are after our attention. Our attention can be defined many ways. I define it as "the second glance." If web marketers or spammers produce something that draws our eye back to it, they have our attention.

Snarkiness will get my attention long enough to accomplish your short-term goal, but it does so at the expense of the long term. If you are snarky to me, I want to listen to you less in the future. You made your point (congratulations), but you did so at the expense of all future points you wish to make with me.

Is that a win for you? (Seriously?)

Putting it into farming terms: Snarkiness is equivalent to eating seeds stored for planting. You are eating today, but you will starve in the future without seeds to plant. And you're starting a negative cycle because without seeds to plant, you will not produce more edible (and planting) seeds.

The Solution

Don't be snarky. To give another Star Trek reference, Wil Wheaton puts the final call about snarkiness succinctly, in what has become known as Wheaton's Law: "Don't be a dick."

Snarkiness may be fun, but it is expensive fun. And the snark foots the bill in terms of his or her own influence and reputation. Is it worth it?

Chapter 47. Question

Question: How many calls from work are too many while you're on vacation?

Answer: One.

This chapter is about how to quickly and efficiently get the employees you need most to begin seeking employment elsewhere.

Chapter 48. Managing Confidence

Hot Chicks - Baby chickens beneath a warming lamp...

This chapter is about inspiring others. My family raises chickens that lay eggs. These chickens are referred to as "laying hens." Natural attrition has taken our flock of laying hens to 11, plus one rooster. We recently received an order of new chicks (pictured above). We keep them inside for the first couple weeks until they grow enough feathers to withstand outside temperatures. The temperatures in our area this winter have been very mild, so we ordered 33 birds earlier than usual. For now, they live in our sunroom in a large box beneath a heat lamp. Hence, they are hot chicks.

As we approach the time when they are ready to be transferred outside, the birds begin to fly. Believe it or not, these chicks can fly a couple feet up, even now. It's funny to watch because they are a little clumsy and their weight isn't yet distributed well for flight. Mostly they lack confidence that they can fly well. How do I know? Because the instant one bird flies out, they all do.

Why is that? They *see* that it is possible. With two cats in the house, we take great care to keep the chicks inside the box. The cats will not bother them while the chicks are in the box, but all bets are off if the birds are running around the house on the floor. I fully expect the cats would chase and catch them.

Right now, the chicks don't know it, but they are capable of escaping the box. So how do we keep the chicks from realizing they can fly out? We manage their confidence. We have a pretty high box, to begin with. They could fly out. We know this. In fact, one flew to the edge of the former box just the other day, prompting a transfer to an even higher box. Once they know they can make the upper edge, we have to change boxes.

Managing Confidence

The reason I'm talking about my hot chicks is that I actively manage their confidence, and I think managers also manage the confidence of their employees. If you manage people, you are managing their confidence, whether you realize this or not. Your team is leaving work each day either more confident or less confident than when they arrived. Stasis is possible, but not likely. Everyone likes to do a good job. And everyone drops the ball at some time or other.

Managing confidence well means you manage each situation individually.

"But that's a lot more work, Andy."

Yes. Yes it is. One of the reasons a manager is paid more is because the job is *supposed* to be more work. Establishing and enforcing blanket rules is not only lazy, but it means you treat your employees like they're in kindergarten.

Do you find yourself complaining your employees behave like unmotivated children? That's a clue that you're treating them like children. Treating your team poorly diminishes confidence. It's offensive personally and professionally.

But that's not your only option as a manager. You can inspire confidence in your team members by treating them with trust and respect. Believe it or not, this is *easier* to manage than demotivating your employees. It's less work in the long run.

How do you inspire confidence in your team? Treat them with trust and respect. Those are broad and vague terms.

Yes, so here is a specific example. Suppose you are asked to provide an estimate for something that's impossible to predict. For example, you need to predict how long it will take your team to figure out the best and fastest way to accomplish this task."

Your first response as a manager should be along the lines of: "It is impossible to know the answer to that question. As a result, everything else I say should be considered at something less than 50% confidence."

If someone wants a number, explain why that's a risk-laden question, and explain the risks. Stand up for your team in these matters.

Sometimes, no matter what you do, you are stuck with an impossible deadline. Years of experience developing software and delivering solutions has taught me that everything is mutable except the delivery date. When faced with those scenarios, I tell my team, "I don't think this is a fair deadline for this project, and I have communicated that fact back up the chain of command. I believe if anyone can bring this in, it's you. And if you cannot bring this in by the deadline, it simply cannot be done."

Faking this doesn't work. It's not an incantation. It's a reminder of the confidence I have demonstrated time and time again in the team. Bolstering the confidence of your team is part of the job of every leader. You want your team members to fly. And if you can encourage the confidence for your team members to succeed, those team members will give you the last percent.

Chapter 49. Push the Pebble

This chapter is about starting something. Today is the first day of something. Somewhere, someone is starting something that will become big. It will impact lives. It will change things, forever. Somewhere else, someone is improving the thing they started recently. They are tweaking, tinkering, thinking, and doing. Is either of these people you? If not, why not?

Dams and Avalanches

Obstacles occur. They are as natural as gravity; they are part of entropy. They block. But they also support. It really depends which side of the avalanche or dam you find yourself on, and your response to it.

For example, if you are leading a convoy or group of travelers and you need to get from Snowy Point A to Snowy Point B, the banked snow between those points may create a stable bridge that allows you to cross safely and quickly. If you enjoy the lake (or power or fresh water capacity) created by the dam, it is a good thing. But if you find yourself beneath an impending avalanche or in a water-restricted area downstream from the dam, you may feel altogether different about them.

From a physics standpoint, both dams and avalanches represent something called potential. *Potential* is stored energy. It is ready to be unleashed for good or harm and is being held back by some force or combination of forces.

One Pebble

Do you see metaphors for dams and avalanches in life and work? Is there something that needs to happen? Some energy that could be released for good? How does such energy get released? Someone, somewhere, starts something.

Avalanches begin when the smallest bit of snow begins moving. Dams fail, beginning with a tiny crack, or with one small pebble becoming dislodged. Once started, all that potential, all that stored energy, begins to work together. If stones and snow were conscious, I doubt the first to move, the *starters*, would realize what they were starting.

Kick the Pebble

Be a starter.

Awesome idea. But where? Where are you right now? Start there. Something needs to get done right where you are. Jeremiah said it best: "If you cannot change where you work, change where you work."

Quit waiting for someone to do something. You do it. And please hurry. The world is waiting.

Chapter 50. Less-Useful Soft Skills

Over a career that spans decades, you encounter useful and less-useful soft skills in the modern enterprise. I thought I would share a few of the less useful variety.

Free Advice

If someone asks you for advice, that's a cool compliment. The person asking has seen something that compels him or her to seek information about how you do or see things. Your perspective is requested, wanted, and welcomed.

That's different from someone offering unsolicited advice. Way different. In the first case, the individual asking is open to receive advice. This is often not the case with unsolicited advice.

I've learned it's best to offer advice only when asked.

Retentiveness

Every field has its standards, and learning them is important. But each person has methods and interpretations of how to make those standards work for them. People's way of doing things needs to be respected.

For example, in the software development world, best practices, formatting, case, and capitalization are all excellent tools for assisting a developer to represent, support, and facilitate their thoughts and thought processes—until the someone decides to transfer a particular interpretation of how to do things to another developer.

I hope you are seated before you read this next sentence: Not everyone thinks like you. (I know!) Moreover, not everyone *wants* to think like you. (Shocker.) Others think in ways that facilitate their code development style. Their documentation—or lack thereof—is there (or not) because of their coding style. Some developers wait until the end of a development project to begin code optimization (this is a recommended best practice, by the way) instead of optimizing each code fragment.

Everyone—and I mean everyone—has a preferred method for representing code or doing whatever their job is. I hear you thinking, "But their way doesn't work for me." It's. Not. Supposed. To. Work. For. You. It's supposed to work *for them*. Different is not wrong.

Allow others to work in their style. If they observe your working style and want your advice, then you're free to give it. Otherwise, don't.

Destructive Competition

Wanting to be a better developer is a good thing. Wanting to be better than the developer in the next cubicle is not a good thing. The collective IQ of a team is greater than the collected IQs of the team members. Why? Synergy is the great entropy-buster, spawning positive cycles as team members interact—*sans* retentiveness. Synergy works best with a dash of humility, and humility facilitates an environment where team members freely seek the advice of each other. Communicating you are the smartest person on the team accomplishes the opposite: striking lines of communication, binding synergy, and destroying the cooperation that would induce a positive cycle. (This is also why I don't like performance-based management, by the way.)

Before acting or speaking, ask yourself, "Will this help?" (And don't assume that just because you want to say it, it will help.) If the answer is not, "Yes, this will help," don't act or speak.

Chapter 51. Can You See Me think?

Once upon a time I was a manufacturing systems integrator. That's a fancy description of a person who designs and builds machine control systems. I was asked to replace a control system and given a tight timeline to accomplish the work. My engineering spidey-senses were tingling, but there were bills to pay and the promise of a large follow-up gig with the same company if I succeeded. So I took the gig.

One of the managers decided to keep me company as I implemented the solution. The other people he supervised did manual labor, so I understood his desire to keep an eye on me. Have you ever watched someone program Programmable Logic Controllers (PLCs) using ladder logic? It's the coding equivalent of watching paint peel off walls. More to the point, it's *nothing* like watching people perform manual labor.

After several long days, I asked this manager, "Can you see me think?" A rather blunt conversation ensued that identified trust as the root cause of his direct supervision. He'd had a bad experience with another manufacturing systems integrator and was determined not to repeat that experience.

Although I implemented the new control system successfully, I was not successful in this role. Hindsight is 20/20. I now realize I managed that conversation poorly. One result was I did not get the large follow-up gig. Another result was I learned some things. I'll share my lessons with you for considerably less than they cost me.

Measuring Manufacturing Work

If you work in a factory, the manager can watch you pack widgets from the end of a manufacturing line into shipping crates. This work is easy to measure. The widgets can be in one of a small number of states: on the manufacturing line, in transit from the manufacturing line to the shipping crate, or in the shipping crate.

This is not the case with work that requires analysis, judgment, and the ability to determine the best solution. Intellectual work is harder to measure.

Measuring Intellectual Work

It is possible to measure intellectual work, just not effectively. Today, you can effectively and affordably measure the *outcome* of intellectual work but not the inner workings of the mind itself. (Note: this will change one day—probably sooner than any of us want. When it does change, I believe it will redefine economics and culture.) Until that change comes, only one metric matters: shipping, delivering, executing.

Please Pay Attention

Intellectual work is art. Software development, for example, is the art of translating business requirements into logic, and then logic into machine language via third-, fourth-, or fifth-generation languages that then translate the code (art) developed by the developer into machine language. However, not everyone understands what that means.

Divide and Conquer?

Modern project management methodologies purport art can be divided into smaller segments and that the segments can be estimated and tallied into a sum of parts. Let's apply that assertion to the art of painting. Let's have an imaginary conversation about one of the most famous paintings of all time: the Mona Lisa.

Imagine you are the project manager hired by Francesco del Giocondo for delivering the painting of his wife, Lisa del Giocondo. Where do you start?

You may begin by asking the painter Leonardo da Vinci for an estimate of how long it will take to complete the painting. After all, a popular theory is the painting was commissioned for a new home and as a celebration of the birth of the del Giocondo's second child. You would assume a timely delivery would be part of the project's commission.

"It's 1503, Mr. da Vinci. How long will it take to create this painting?" you ask.

"I don't know," he replies.

"Mr. da Vinci, I do not understand. The reason you have been commissioned instead of your competition is because you are one of the best. This is not your first painting, sir. Surely you can base an estimate on how long you've spent on paintings of similar size and complexity?"

"All paintings are different," da Vinci replies, "and the differences are not always apparent to people who do not paint."

So you offer to help by applying principles gleaned from YOPMC (Ye Olde Project Management Certification): "There are phases, right? First you prepare the canvas, then draw an outline of the foreground, then paint the foreground, then outline the background, then paint the background, correct? You may not do it all in that order, but that's what you do, isn't it?"

Da Vinci replies, "Those are the mechanics, yes. There is nuance that goes well beyond what you describe. I can do everything you stated and produce rubbish."

"Still, though, we can break the project into these pieces and guess at how long each will take, right? We won't hold you to these estimates, of course. We're just attempting to get a rough estimate so we'll know the month we can plan a party to unveil your painting in our new home."

The Mona Lisa was completed in 1517. That's 14 years after it was commissioned. As an artist, was Da Vinci satisfied with the result?

"Leonardo, later in his life, is said to have regretted 'never having completed a single work'." – from the <u>WikiPedia</u> article: <u>Mona Lisa</u>

The Point?

In the context of measuring intellectual work, this imaginary dialog makes several points:

- It is always possible to measure the outcome as a single point of reference
- It is not always possible to achieve a goal by dividing a thing into its constituents and then achieving the pieces
- Even when you have all the requirements, accurate estimation is difficult
- "I don't know" is the best thing to say when it is true

The primary point is you cannot see me think. Therefore you cannot directly measure the efficiency of my intellectual work. You are limited to measuring the outcome of my thinking, the delivered solution. Therefore, the result is all you should measure.

Made in the USA
Lexington, KY
08 February 2015